BUILDING AUTOMATION SYSTEMS A TO Z

BUILDING AUTOMATION SYSTEMS A TO Z

HOW TO SURVIVE IN A WORLD FULL OF BAS

Phil Zito

Published by Building Automation Monthly

Building Automation Monthly is the premier training website for Building Automation Systems. It is the mission of Building Automation Monthly to equip 20,000 new building automation professionals by the year 2025.

ISBN-13: 9781539914488
ISBN-10: 1539914488
Library of Congress Control Number: 2016918603
CreateSpace Independent Publishing Platform
North Charleston, South Carolina

DEDICATION

To my wife Rachel and my three children Gabby, Trevor, Audrey.

Thank you for supporting me during the long nights when this book was just another one of my crazy ideas.

BUILDING AUTOMATION MONTHLY MISSION STATEMENT

To provide the best online Building Automation training to prepare 20,000 people to enter the Building Automation space by 2025

TABLE OF CONTENTS

ACKNOWLEDGEMENTS

Without my 79 pre-sale customers this book would not be here today. Before this book was anything more than an idea you all decided to take a chance on me.

Through your feedback, encouragement, and motivation I was able to create a book that will change the industry. Never before has all of this information been collected. I have each and every one of you to thank for this.

To my friends at Control Trends, Eric, and Kenny, thank you so much for encouraging me to pursue my dream of teaching building automation!

To my mentor and friend Keith Reller, the best sales person and best person I have ever known, thank you for showing me how to let my "personality" out :-D.

And last but not least, thank you to all of you in the BAM Nation. I appreciate all the kind words and encouragement as I went down this path!

As I mentioned, this book would not be here without the financial support of my 79 pre-sale customers. Here are the names of those who included their name in the pre-sale order form:

1. Adam Lazare
2. Adam McGuire
3. Aditya Prabhu
4. Adrian Altine
5. Amir Shabani
6. Andrew Collins
7. Andy Schonberger
8. Anthony Borla
9. Brian Graham
10. Bruce Duyshart

11. Chad Chenier
12. Christian DeFehr
13. Cornell Wharton
14. Darragh Gleeson
15. David Katz
16. Derek McGarry
17. Ed Selkow
18. Eddie Qualls
19. Fernando Palos
20. Fiona Murray
21. Golan Dachoach
22. Greg Cmar
23. Greg Galusha
24. Jeff Woolums
25. Jeremy Leahy
26. Jesse Schwakoff
27. Johannes Van Haren
28. John Fox
29. John Guy
30. John Kostrzewa
31. John Tegan
32. John Zampetti
33. Jonathan Stahl
34. Jonny Fernández
35. Jorge Martinez
36. Joseph Ament
37. Juan José Castillo Fernández
38. Justin Dixon
39. Ken Smyers
40. Kent Shasta
41. Kyle Shipp
42. Larry Schoeneman
43. Larry Stevens
44. Lauren Marotta
45. Lawrence Reeves
46. Lee Bridgeford
47. Marek Kozlowski
48. Mark Mueller
49. Mayuresh Kulkarni
50. Meshack Otedo
51. Mike Case

52. Mike Scott
53. Mike Zilm
54. Niel Carter
55. Patrick Cahill
56. Paul Disher
57. Paul O'Connell
58. Peter Bell
59. Raniel Camacho
60. Rich Mccorkle
61. Richard Bluestone
62. Rob Folger
63. Rob Stewart
64. Robert Nuckols
65. Robert Turner
66. Sandeep Kulkarni
67. Scott Drabek
68. Scott Herman
69. Sheldon Gabriel
70. Stephen Hennessy
71. Steve Jones
72. Tom Edsall
73. Tony Poland
74. Tyson Sargent
75. Valentin Berrelleza
76. Vikram Bhardwaj

INTRODUCTION

Thank you for purchasing this book!

I spent the second half of 2016 pouring my heart and soul into this book. It is my hope, rather my mission that this book will impact your career more than any other training product you have ever invested in.

How did this book come about?

It all started back in September of 2012 when I decided to start Building Automation Monthly. At the time, I was spending hours writing training documentation for my job.

I had told my wife how energized I felt when I taught others. My wife, being the amazingly insightful woman that she is, suggested that I take my teaching beyond just my company and start writing a "blog" on Building Automation Systems (BAS).

At that moment Building Automation Monthly was born.

During my first year of blogging, I struggled with doubts.

I often asked myself who am I to be teaching folks?

I went through a period where I felt like an imposter. I'm sure that came across in my early posts.

Have you ever felt that way?

Like you are just pretending and waiting to be found out?

I struggled with this for a period of time, but as I began to write my articles, I started to get e-mails from folks telling me how my teaching had changed their careers.

What made my day was when folks told me that the way I explain complex topics made them "click."

After blogging for four years on all things BAS I had a light-bulb moment.

I was struggling to keep up with all of the questions that were getting e-mailed to me on a daily basis. With a full-time job, managing the Technical Integration Program for a Fortune 100 BAS company, I just didn't have the time to answer everyone's questions.

So, I came up with an idea.

I decided I would gather information from the internet to answer all the questions I had been getting.

But there was a problem.

I quickly realized that there was very little content available for me to reference. I mean, of course, there was "content" out there, but this content was just sales letters and marketing fluff tucked into fancy white papers.

I spent two years looking for information. I mean surely there was something out there that could teach folks the foundational knowledge they needed to understand and manage a BAS, right?

The book is born

So, search I did, forums, books, articles, etc.

I read, read, and read some more…

There were a few good books on Smart Buildings, Energy Management, and HVAC but, no matter how hard I looked, I couldn't find a solid primer on Building Automation Systems. It was then, sitting at my desk in April 2016 that the idea for this book was born.

But how do you write a book?

As I would later find out, the normal process for writing a book is that you come up with an idea, submit that idea to a publisher and then work to create the narrative (that's what publishers call the text that ultimately becomes a book).

Well, I didn't go that route.

I thought to myself, what if, I could gather a bunch of professionals in the building automation space and have them give me feedback as I wrote the book.

What if I could tap into the knowledge of architects, contractors, programmers, facility directors, and many other professionals, instead of just relying on my wisdom and experience?

I found out rather quickly that no-one had taken this approach before. So July 2016, 79 other professionals and I began the journey of creating the book you are reading today.

Over the period of 4 months, I produced chapter after chapter and submitted it to this group for review.

The diversity of the group's feedback was invaluable in helping me to think of building automation from multiple angles.

Ultimately, the result of this "crowd-sourced book, is the ultimate building automation starter guide.

I want you to look at this book as your mentor in a box. This book will give you the fundamental knowledge of how building automation systems are installed, managed, and supported within the built environment.

How is this book structured?

Let's face it, most books that teach you a technical topic are great for curing insomnia. Unfortunately, they're not so great at keeping the reader engaged in the learning process.

Well, my readers, you are in luck. This book is jam-packed with stories and humor that will keep you engaged while you learn the fundamentals of building automation systems.

So, how exactly am I going to take you through the learning process?

This book is broken out into five sections. These sections are designed to build upon one another and you should read them in order.

Each section contains several chapters. These chapters begin with an overview of the topic and a list of the subtopics that I will discuss. Each chapter will conclude with a summary of the chapter.

My goal is that you will be able to apply the knowledge you gained from each chapter immediately.

When you follow this book from start to finish your knowledge of building, automation systems will put you in the top 10% of all building professionals!

That's a large claim for me to make, but I believe by the end of this book, you will agree that this is the book on building automation systems that you have been missing throughout your career.

I want to get you into the book, so I won't spend a lot of time describing each section in this chapter. Here's what you can expect from each section.

Section 1: Core knowledge

Section 1 provides you core knowledge you are going to need to have a successful career in the building automation space. You will gain the baseline knowledge you need no matter what your skill level is coming into this book.

The section consists of six chapters that will build upon one another. Once you are done reading these chapters, you will be prepared to apply what I teach throughout the rest of the book. It is important that you read through these chapters in order.

Chapter 1: HVAC basics

It all starts with HVAC. Without Heating Ventilation and Air-Conditioning, the building automation system has nothing to control. In chapter 1 I walk you through the fundamentals of HVAC.

Chapter 2: BAS fundamentals revisited

In chapter 2, I teach you the fundamentals of building automation systems.

Chapter 3: Smart building systems

In chapter 3, I show you to the most common systems that you will encounter during your building automation career.

Chapter 4: IT fundamentals

In chapter 4, I walk you through the fundamentals of information technology. I teach you how information technology works and how information technology groups are structured.

Chapter 5: Electrical fundamentals and energy management

Possessing an understanding of how electrical systems work and how energy is managed is critical for anyone in the building automation space. I discussed these exact topics in chapter 5.

Chapter 6: Standards and organizations

It's important to understand the standards and organizations that exist to support building automation systems. In chapter 6, I show you the standards you need to understand and the organizations you should connect with.

Section 2: The art of buying, installing and upgrading a BAS

Now that you have the fundamentals down, it's time to learn how to buy, install and upgrade a building automation system.

Chapter 7: Procurement

No matter who you are, or who you work for, at some point in your career, you will be involved in the purchase of a building automation system. Even if you do not directly approve the purchase of building automation systems, you can still influence their selection process. In chapter 7 I will explore the procurement process.

Chapter 8: Construction

Before there is a building, there is a construction project. In chapter 8, I discuss how the construction process happens and what that means for building automation professionals.

Chapter 9: Upgrading the BAS

Nothing last forever, right? If you stay with the same company for any amount of time, you are bound to be involved in an upgrade. In chapter 9, I walk you through how to manage the upgrade process.

Section 3: After the trucks leave supporting your BAS

Once all of the contractors are gone, and the BAS is your own, what do you do?

How do you manage the building automation system?

I will discuss these topics and more in this section.

Chapter 10: Managing a BAS

How can you manage a BAS?

What do you need to manage?

What actions can you take to make your BAS as effective as possible?

I answer these questions in chapter 10.

Chapter 11: Managing service providers

Managing service providers is a challenge for many building automation professionals.

How do you know when you need a service provider?

Can you self-perform your maintenance?

What maintenance tasks would those be?

I answer these questions in chapter 11.

Chapter 12: Advanced maintenance management

In Chapter 12 I cover advanced topics related to maintenance management.

Section 4: Advanced topics analytics, IoT, and integration

In Section 4 I move into the advanced topics that are impacting the building automation space. In this section, I will teach you what these topics are and how to implement these topics.

Chapter 13: Analytics

Analytics are all the rage right now!

In Chapter 13, I walk you through what analytics are and what they are not. I then run you through the three most common scenarios that analytics are used for. I close out Chapter 13 by giving you a step-by-step process for implementing analytics.

Chapter 14: Internet of Things

The Internet of Things (IoT) is a confusing term for most people. That's why in chapter 14 I help you understand what IoT is and where and how it applies to your building automation system.

Chapter 15: Systems integration

Systems integration is becoming quite common. Designers and engineers are beginning to combine building systems to drive greater outcomes. Naturally, this requires the integration of systems. In chapter 15, I discuss systems integration and lay out my step-by-step process for approaching this complex topic.

Section 5: Concluding Thoughts

In this section, I lay out a path for you to continue growing your career I also discuss the multiple different ways in which you can use this information to take your career to the next level.

Chapter 16: Next Steps

The end is just the beginning. By the time you reach chapter 16, you will have a level of knowledge around building automation systems that will prepare you for whatever path you decide to take in your building automation career. I will also discuss the different paths you can take to continue with your building automation career.

The choice

There you have it, folks, that is what you can expect from this book.

Now, I'm going to give you a choice.

Some of you may be tempted to just skim this book. That's fine. I encourage you to use this book as a reference. However, you will not get the most out of this book if you chose that path.

For the rest of you, I encourage you to embrace each chapter.

Read through the chapter once for understanding and then a second time for comprehension.

No matter which path you choose, I welcome you to the Building Automation Monthly (BAM) Nation!

Now, let's get your learn on!

SECTION I – CORE KNOWLEDGE

Section Overview

For some of you, this may be your first exposure to the world of building automation. If that is the case, then let me be the first to welcome you to the crazy world of building automation!

Even if this is not your first rodeo, I believe you will find the information I teach you in this section useful, after all sometimes we forget the things we haven't used in awhile…

In this section, I will be taking you through 6 chapters. These chapters are designed to build upon one another, so I encourage you to go through the chapters in order.

With that being said, let's move into the first chapter.

CHAPTER 1

HVAC BASICS

What's in this chapter?

B efore I even get into building automation systems, information technology, and all that other fun stuff, I need to cover the fundamentals.

The most fundamental topic in the BAS space is the topic of heating ventilation and air-conditioning, also known as HVAC. To really understand what building automation systems do, you need to have a fundamental understanding of the HVAC systems they control.

That's why in this chapter, I will be discussing the topic of HVAC!

This chapter will cover the four main areas of HVAC fundamentals. Those four areas are:

- HVAC systems to including airside systems and waterside systems
- HVAC components
- The refrigeration cycle
- Psychrometrics

By the end of this chapter, you will understand how heating, ventilation, and cooling works within your building(s).

Now, to some folks HVAC may seem like a complex topic. However, I promise you I will make HVAC fun and easy for you to understand. You'll walk away from this chapter understanding how the systems inside your building work.

So if you're ready, to join me as I dive into our first topic!

What is HVAC

Heating ventilation and cooling, also known as HVAC, is a very important topic.

I grew up in Houston and Houston is ridiculously hot.

At the time of me writing this chapter, I'm in Milwaukee Wisconsin, and it's the middle of summer. Lately, it's been getting up to 90° and folks are complaining about the heat. This makes me chuckle, because when I was growing up in Houston, 90 ° was cool!

You see a little-known fact is that; Houston was built on a swamp. This makes Houston quite humid.

Do you know what made that hot, humid swamp bearable?

It was the fact that at any time I could go inside my house and have some nice ice-cold air conditioning.

But it wasn't always that way.

Heating ventilation and cooling is relatively new. While I won't get into how air-conditioning was invented, I am going to explain the basic systems that you will find in a traditional building.

Over my many years of teaching and working in the BAS/HVAC world, I've found that the best way to break out HVAC systems is to divide them into airside and waterside systems. I'll explain to you what each of those systems are in the next section.

Airside systems

Airside systems move air. Air is the primary method to heat and cool spaces within a building. In this section, I am going to describe the most common airside systems.

Rooftop units

Rooftop units are one of the most common types of air distribution systems.

Often I get asked, "Phil, what is the difference between an air handler and a rooftop unit"?

To take the average person an air handling unit and a rooftop unit look pretty much the same.

At first glance, they also function the same way. But there're a couple of key differences between air handlers and rooftop units.

The big difference between an air handling unit and a rooftop unit is a rooftop unit is designed to be outside. Hence the name rooftop unit.

Another big difference is that many rooftop units are self-sufficient.

Take a moment to think about an air handler. Now I know I haven't covered air handling units yet, but that's okay.

One of the most distinct differences between an air handling unit and a rooftop unit is that an air handling unit usually requires heating and cooling from other sources.

When I reference other "heating and cooling sources" I am referencing what we in the industry call waterside systems.

Now, I realize some air handling units support direct expansion cooling, and some rooftop units support chilled water coils. This distinction is not true 100% of the time, but it is mostly true.

Typically, when you purchase a rooftop unit, you are purchasing a self-contained unit. A self-contained unit has the heating and cooling built into it.

Oftentimes a rooftop unit will have what is called DX cooling. DX cooling stands for direct expansion. DX cooling uses refrigerant to cool the air. DX cooling works by moving the refrigerant inside the system through a series of mechanisms that I will cover in the refrigeration cycle section.

Ultimately, what you need to know is that the **DX cooling coil will absorb the heat from the air** inside the unit.

A rooftop unit will typically have a heating mechanism, cooling mechanism, and a supply fan.

Some units will include things like humidifiers, enthalpy wheels, dampers, and variable frequency drives (VFD's). Typically, with rooftop units, whatever you need is built into the unit. This allows the unit to be independent of any other system.

This kind of system is good for commercial office buildings or schools that may not have central utility plants.

Rooftop units are sized according to the amount of air, measured in cubic feet per minute (CFM), that they need to provide and the amount of cooling capacity, measured in tons of cooling, they contain.

Cooling capacity is indicated by a unit of measurement known as a "Ton of cooling." A **ton of cooling provides** the unit with the ability to remove 12,000 BTUs (British Thermal Units) per hour. Historically a ton of cooling meant the amount of heat removed by the rooftop unit would melt a ton of ice in an hour.

That's some historical trivia for you right there!

Air handling units

Air handling units (AHU) are different from rooftop units.

An **air handling unit** is a unit that typically, is **not** independent of the primary heating and cooling system and is usually used inside a building.

Air handling units are often custom built on-site or delivered in sections that are put together at the site. Inside an AHU you have several systems. Typically, an AHU will have a Fan, Cooling and heating coil, dampers, actuators, valves, and sensors.

Later in the chapter, I will break out what each of these pieces are.

An AHU operates by moving air into the building via ductwork. Inside the AHU the air is heated or cooled via a series of coils. Usually, these coils are supplied hot or chilled water from the waterside systems.

One of the potential downsides with AHU's is that if the waterside systems go multiple air handlers can be affected.

If a self-contained rooftop unit were to lose its ability to cool, only affect that one unit would be affected.

One reason why you would want to use AHU's instead of rooftop units is that rooftop units can introduce energy inefficiencies due to their design.

This is because waterside systems, which typically supply air handlers, are more efficient at producing heating and cooling then self-contained rooftop units.

There are two kinds of AHU's you will encounter in buildings. Constant and Variable Air Volume (VAV) units.

But before I go there I'm going to take a second to unpack a term you've heard throughout this chapter.

Air volume

Throughout this chapter, I've used the term air volume, and some of you may be wondering what that term means.

Here's how the ventilation side of HVAC works.

Air volume is one of the concepts that is little difficult for folks to grasp at first because it's hard to see air.

Believe it or not, air has volume. This means that air takes up space. Now that may be a hard concept to grasp because you can't see air.

So follow along as I take you through a little bit of a journey.

This journey begins with a pool.

This pool contains water, and you can only put so much water in a pool. If you put too much water in a pool, it will overflow.

Air acts the same way. You can only put so much air in a space before that air wants to escape. It is this principle, called **pressure** that allows air to be moved throughout a building.

An air handler provides a volume of air which is typically measured in cubic feet per minute (CFM). This air volume is being provided to a space or series of spaces. When this new volume of air is supplied to the space, the air that is currently in the space is exhausted out of the room.

This concept called an **air change** is when the full volume of a space is replaced. An air change is a critical process to ensure that a space is heated and cooled.

This is achieved by exhausting the same amount of air that enters a space. The amount of air changes is calculated by how many times the volume of air in a room is replaced within an hour.

Constant volume systems

The first type of air handling unit I will cover is the constant volume unit.

A **constant volume unit** provides a constant amount of airflow into your spaces.

Usually, the control sequence for a constant volume unit requires you to reset the discharge air temperature set point based on either the return temperature or the average space temperature. Since the constant volume unit provides a constant CFM flow you cannot use some of the pressure control strategies, you will see in the next section.

Variable air volume (VAV) units

The second type of air handling unit is the variable air volume unit.

The variable air volume unit often called a VAV unit is a very common unit and can be found in most medium to large buildings. The VAV unit addresses one of the biggest downsides of the constant volume unit, constant airflow.

The VAV unit varies the airflow based on the systems pressure requirements. A pressure sensor measures the systems pressure. Often you will hear this sensor called the differential pressure sensor or $2/3^{rd}$ sensor (meaning 2/3rds of the way down the longest duct).

You only need to change the air within a space as required to meet the temperature, humidity, and C02 levels.

Side note: You will learn more about airflow design for specific spaces when I review ASHRAE's 62.1 2016 standard in chapter 6.

Now that you have control of the airflow, you are no longer subjected to air temperature modifications being your only form of control.

You now have the ability to modify airflow instead of temperature.

While there are plenty of benefits of variable air volume units, there is one capability of variable air volume units that you need to be aware of. This capability is known as static pressure control. When you begin to vary the volume of air inside the ductwork, you need to be aware of the impact that this increase or decrease of air pressure will have.

This increase or decrease in air volume is called static pressure.

Static pressure

When it comes to air, there are three static pressure control scenarios.

Duct static pressure is the first type of pressure control. **Duct static pressure**, is the amount of air pressure measured in inches water column (in. WC) that is present within the ductwork. As an air handler increases its airflow, the volume of air will naturally increase inside the ducts. This will result in an increase in static pressure. If the static pressure becomes too high, the ductwork could become damaged or destroyed.

The second type of pressure control is building static pressure. Building static pressure is also measured in inches water column. However, building static pressure measures the amount of air pressure within the building.

The third type of pressure control is space pressure control. Depending on the space type you will want to keep the space pressure positive or negative.

When a space has positive air pressure, that means the air is being pushed out of the room. This is good for clean rooms, and other spaces that do not want outside particulates within the space.

The opposite of positive air pressure is negative air pressure. Negative air pressure is when the air from outside the space is being pulled into the space. This is good for isolation rooms, where you may not want the air within the room to leave that room. Static pressure is a direct result of the amount of volume being put into a specific space or piece of ductwork.

There are many ways that static pressure can be increased or decreased.

The primary way is to utilize what is known as a variable speed or frequency drive. A variable speed or frequency drive allow the fan on the variable air volume unit to increase or decrease the amount of air volume that it provides.

A VFD or VSD works by increasing or decreasing the electrical signal to the fan. This, in turn, causes the motor to produce a stronger or weaker electronic field. It is this field, or force as some folks like to call it, that increases or decreases the motors rotations per minute or RPM.

The second way that static pressure can be increased or decreased is through the use of dampers. Dampers allow you to block or divert the flow of air. Air can be allowed to exhaust from or bypass to specific parts of the building or ductwork. This will allow you to increase or decrease the pressure.

Bypass dampers are a common method that constant volume units use to handle variable air flow. Please be aware that using bypass dampers instead of a variable air volume unit, is not recommended.

Terminal units

Terminal units, often called Variable Air Volume (VAV) boxes, work by varying the air supplied to the space. Some terminal units provide constant air volume to spaces. These terminal units are called Constant Air Volume (CAV) boxes.

VAV units are used to vary the airflow into a space. Now as you may recall, the temperature in a space is changed varying the amount of conditioned air into the space.

Here is how this works.

A terminal unit will typically have three airflow set points. These set points are:

- Cooling flow
- Heating flow
- Cooling max flow

These set points allow the box to manipulate the temperatures in the space by replacing the volume of air in the space with warmer or cooler air.

When a box needs to provide more cooling, it will increase the amount of airflow it provides.

On the flip side, when the box needs to provide heat, it will reduce the airflow. The decrease in airflow during heating serves a purpose. A slower moving mass of air can exchange temperature more efficiently.

Since the cooling is cooled at the air handler and the air is heated at the box you'll often see the VAV box airflow reduce when you're in heating mode.

It's important to remember that slower air will allow the heat to transfer to the air more easily.

To visualize what I am talking about, imagine you are putting your bread in a rotary toaster.

At my work they have this rotary toaster that I put my bread on. Even though I can set the temperature and the speed, I always burn my bread.

I can figure out computer programming, energy management, and cyber security but I cannot figure out how to make this darn rotary toaster work!

Ok, what was I talking about?

Oh, ya, heating.

So, just like that rotary toaster, the slower your airflow, the more heat will be transferred.

The BAS will change the airflow of the terminal units to control the space temperature

Up to this point, I have talked about air systems, but I haven't talked about how these air systems are "calibrated."

In the construction chapter, I cover a concept called balancing. **Balancing** is the action that makes the BAS "accurate." In this next section, I am going to take a brief segue to talk through this concept.

Balancing

Balancing is an important concept for you to understand.

Back in 2010, I was living in Dallas. My new house had just been built, and I was excited to have my own office. This new office was going to give me a space where I could work from home in relative peace and quiet.

However, when the summer heat started to kick in, I quickly realized that there was a problem with my office.

There were two issues with my office. The first issue was that I was not getting enough airflow into my office. The second issue was that when I shut the doors to my office, the HVAC system was not able to exhaust the air that was in my office. It turned out that this was because the main exhaust vent was located outside my office.

I had the folks who installed the system come to my house and install a new diffuser. While that did allow more air into the office, it did not resolve the issue of exhausting the hot air from my office.

Looking back, I realize, that even in residential spaces, the proper balancing of your diffusers and exhaust vents is critical for ensuring that your HVAC performs properly.

The thing is, a fair amount of folks who work within the building automation space don't pay attention to balancing. They simply sit alongside the balancer and enter the calibration set-points for the airside or waterside systems into the BAS while the balancer is balancing the system.

I was fortunate that early on in my career the company I worked for did their own balancing. This meant that I had to go and assist the balancers on my projects. Because of this, I got to learn how to use a balancing hood and other balancing tools.

It is very important for you to understand balancing concepts.

Now, what are those concepts you may ask?

Based on my experience, you need to understand the following concepts:

- Air changes, which fortunately I covered earlier
- Calibration factors (also known as "k factors") which I will cover a little later in the section
- Airflow distribution

Airside systems are not the only systems that are balanced. There is a waterside portion of balancing that includes calibrating valves, and ensuring that you have proper flow throughout your coils. Balancing waterside systems is fairly similar to balancing airside systems.

Airflow starts at the air handler. The air handlers fan(s) is/are designed to supply a certain amount of CFM at full speed.

To balance a system, the balancer will turn the air handlers fan speed to 100% and open all of the VAV boxes. Once the fan is set to full speed and all of the boxes are open to hundred percent, the balancer will adjust the balancing dampers to ensure that the diffusers supply the designed CFM.

The next step is for the balancer to measure the CFM at the diffuser(s) for each VAV box. Sometimes there are multiple diffusers for a single VAV box.

In this situation, the airflow from the diffusers is totalized, and the total airflow is compared to the airflow reading in the building automation system (BAS) controller. The BAS controller's airflow reading is then adjusted, using a value called the "K factor."

Now, for those of you have done balancing in the past you know that there is much more to this. You'll notice that I didn't discuss box size, I didn't discuss damper type, and other things that could impact air-flow accuracy.

It's not my intent to teach a full-scale balancing course. Rather, I want to teach you the primary aspects of balancing so that you are familiar with the process.

Now back to that thing called K-factor.

What happens next is that the balancer will work with the BAS technician to find out what airflow the BAS is displaying. It's at this point that the balancer and the BAS technician compare their airflow values and modify the K factor to get the values to match up.

By default, the K factor will start at 1. The measured airflow is the reading times the K factor. For example, if you had a reading of 100 CFM at the BAS and a K factor of 1, you would see 100 CFM displayed (100 CFM* 1). If your K factor was 0.9, then you would see 90 CFM displayed (100 CFM * 0.9).

If the balancers airflow measurement is below what the BAS technician sees inside the BAS, the K factor needs to be decreased. Likewise, if the balancers airflow measurement is above what the BAS technician sees the K factor needs to be increased.

Quite simply, the K factor provides a way for the BAS technician, to "tune" the airflow.

Well, there you have it, folks. You now know how most of the air side systems work within a building. You can now comfortably sit in meetings and understand the terminology and discussions that are happening.

With your airside knowledge firmly established I am going to walk you through the next major system type, known as waterside systems.

Waterside systems

As I mentioned earlier in this chapter, there are two types of systems, airside systems and waterside systems.

Waterside systems are used to distribute cooled or heated water throughout a building. Sometimes you will hear waterside systems called hydronic systems the names are interchangeable so don't be confused.

As far as I am concerned, waterside systems break out into two primary system types, chilled water systems, and hot water systems.

In larger buildings, you will tend to encounter waterside systems as the primary source of heating and cooling. This has to do with the fact that water is better at heat absorption and transfer than air.

Chilled water systems

There are two main types of chilled water systems, chilled water systems, and condenser water systems. One could effectively argue that chilled water systems and condenser water systems are one in the same and that they form a complete system.

For the sake of your sanity, I'm going to break up the two.

First off I am going to discuss chilled water systems. Chilled water systems include chillers and their primary and secondary loops.

Chillers, as you can probably guess from their name, chill the water that is supplied throughout the building. I am going to cover the two main types of chillers. These are water cooled chillers and air cooled chillers.

Water cooled chillers

Water cooled chillers, contain refrigerant that allows the heat to transfer from the chilled water return to the condenser water supply, more on this later.

The important thing to note is that water cooled chillers use water to transfer heat from the chilled side of the chiller to the condenser side the chiller.

Sometimes you'll hear the sides of a chiller called the evaporator and condenser side. The evaporator side would be on the chilled loop, and the condenser side would be on the condenser loop.

So what is a condenser loop?

Energy cannot be destroyed it can only be transferred. Some of you may be wondering why I am talking about energy. The reason I am talking about energy is that heat is energy.

With water cooled chillers you're taking the energy out of the chilled water return and transferring it to the condenser water supply. This condenser water, which now contains the heat from the building, makes its way to what is called a cooling tower.

The condenser water will enter the cooling tower, either from the top of the cooling tower or the bottom of the cooling tower, often called the basin. The water then is cooled using a process called evaporative cooling.

Evaporative cooling is where air flows across the water and absorbs heat from the water through the process of evaporation.

I'm not going to give you a chemistry lesson, is that even chemistry or is that thermodynamics?

Who knows, and honestly it doesn't matter for what I'm teaching in this book.

What you need to understand is that heat is taken out of the chilled water and transferred to the condenser water.

The heat in the condenser water is then moved to the cooling tower where the cooling tower evaporates it. The now cooled condenser water returns to the chiller to absorb more heat.

Air cooled chillers

Air cooled chillers use the same evaporative process as the cooling tower to exhaust the heat from their water.

The process to transfer the heat from the evaporative side of the chiller to the condensing side is called the refrigeration cycle. I will discuss the refrigeration cycle in detail a little bit later in this chapter.

Right now it's important to know that refrigerant changes state at a different temperature than water. Here is how I like to explain this process.

Back in school, you were probably taught that water turns to ice at 32° and it boils at 212° (sorry folks no metric system in this book :-D). Those two temperatures are the temperatures at which water changes "state." **State** is the form that matter exists in. Water can exist in 3 different states, solid (ice), liquid (water), and gas (vapor).

Refrigerant can also change state. The big difference is that refrigerant changes state at a different temperature than water. Because of this, you can take advantage of what's called a state change. To change state, matter needs to expend a lot of energy or BTUs. It is this usage of energy that takes the heat out of the water.

Primary and secondary loops

Before I move on to the hot water section, I want to cover primary and secondary loops.

Both chilled water and hot water systems have primary and secondary loops. A loop is a way to circulate water to and from the waterside system.

Primary loops

Primary loops are the primary loop for the waterside system.

Seems pretty obvious, right?

The primary loop supplies water from the waterside system to the secondary loop or airside systems. This loop connects to the Waterside system(s) that provide cooling or heating.

Sometimes this loop will be directly connected to airside systems. However, in larger buildings, the designer will often see use a secondary loop.

Secondary loops

The purpose of a secondary loop is to take water from the primary loop and supply it to the airside systems. This can be done through a variety of methods.

Mixing valves

The mixing valve is one of the most common methods used to transfer water from the primary loop to the secondary loop.

The BAS will utilize a mixing valve to maintain a constant temperature in the secondary loop. The BAS will adjust the mixing valve to introduce water from the primary loop to keep the secondary loop at set point.

The benefit of using a mixing valve is that the system is only supplying the chilled or hot water that the units need. This allows the waterside systems to only produce the heating or cooling that is required by the secondary loop.

Decoupled loops

Another form of secondary loops is called the decoupled loop. The decoupled loop is a complex control mode. With the decoupled loop you are taking advantage of positive and negative flow.

Now, what exactly do I mean by positive and negative flow?

On a variable volume secondary loop (this means the amount of water moved varies), the secondary loop pumps will vary their speed based on how much flow is required. This can result in the secondary loop requiring more GPM than the primary loop is producing.

This will create positive flow on the secondary loop because the GPM of the secondary pumps is higher than the GPM of the primary pumps. This will result in water from the primary loop being pulled into the secondary loop.

As valves begin to close on the secondary loop and the water pressure drops, the secondary loop pumps will begin to slow down. This will result in the GPM on the primary pumps being greater than the GPM of the secondary pumps.

This will now create a positive flow on the primary loop which will draw water into the primary loop.

Even though that was a very simplistic explanation of the decoupled loop, it should help you understand the concept.

Hot water systems

As I mentioned earlier in this section waterside systems are composed of both chilled and hot water systems.

The most common hot water system that you will encounter is known as the boiler. **Boilers** produce heat by consuming a fuel source. In this section, I am going to cover the two main types of boilers. Those two types are steam and hot water boilers.

I am going to cover both of these at a high level.

Steam boilers are often utilized to provide heating in the form of steam to a campus.

As you may recall earlier in this chapter, I mentioned that water could contain more BTUs than air. I also mentioned that a lot of BTUs are required to change state from solid to liquid, to gas.

Each time matter changes from solid to liquid, to gas it is capable of holding more BTUs. Water requires a lot of BTUs to become steam. Because of this steam carries a lot of heating energy.

To give you an idea of how much more energy it takes to change water into steam, consider this. It takes 180 BTUs to take 1lb of water from 32° to 212°.

To change that same 1 lb. of water from 212° water to 1lb of steam takes about 1000 BTUs. That is almost ten times the amount of BTU's.

Often time's campuses will use high-pressure steam to transfer heat from a central heating plant, sometimes called a central utility plant, to a building. Usually, the steam is then reduced to low-pressure steam and transferred to a hot water loop through a device called a heat exchanger.

Now that I've covered the concepts of steam, I'm going to get into what an actual boiler is.

Boilers

As I mentioned early in the chapter, boilers can be used to produce either steam or hot water.

Hot water boilers, like steam boilers, utilize some form of fuel to heat water. Typically, the fuel sources that are used are gas, wood, or electricity. This fuel is then used to produce heat which is transferred to the boilers supply water.

From that point forward the hot water supply loop acts similar to the chilled water supply loops. Just like chilled water loops, hot water loops can have both primary and secondary loops that distribute hot water throughout the building.

HVAC components

Throughout this chapter, I've mentioned several HVAC components, but I haven't defined these components for you. That was by design.

I didn't want you to get so caught up in the individual objects that you missed how the systems work. Now that I've covered the systems at a high level I am going to dive into the individual pieces that make up the systems.

The components I am going to cover are:

- Dampers
- Valves
- Actuators
- Fans
- Coils

Dampers

Dampers exist to block or modulate airflow. There are many forms of dampers, but for the sake of our conversation, I will only talk about round dampers and square dampers.

Round dampers are typically found in VAV boxes and exist to modulate the airflow supplied to a space.

Square dampers tend to exist within air handlers and rooftop units. These dampers also serve to modulate airflow that is exhausted from and supplied to the units.

Actuators control dampers. Actuators are mechanical devices that connect to the damper shaft.

The **damper shaft** connects to the blades of the damper using a damper linkage. When the damper shaft is turned, it pulls on the linkage causing the damper blades to open or shut.

Actuators

Actuators are mechanical devices that exist to modulate both dampers and valves. Actuators are typically powered by 24 volts alternating current (VAC) or 120 VAC. There are three main types of actuators which I discuss in chapter 2. Actuators will typically have a power source and a control signal. This control signal will typically come from the building automation controller.

Actuators are rated in foot-pounds, which measures the amount of pressure the actuator can apply to close or open a valve or damper.

Valves

Valves exist to close off or restrict the flow of a liquid.

Valves are commonly associated with water but can be used to control other liquids as well. There are multiple types of valves and discussing those types are beyond the scope of this book. The important thing for you to realize is that there are specific types of actuators for specific types of valves. These actuators will typically mount on the valve stem.

The **valve stem** is similar to the damper shaft in that it can be rotated to open or shut the valve. It's important to understand the amount of pressure that is flowing through the piping so that you size both the valve and the actuator properly.

Fans

Fans supply and exhaust airflow within the building. There are multiple forms of fans. Discussing these forms of fans is beyond the scope of this book. The important thing to know is that a motor powers a fan and this motor will cause the fan to rotate.

Earlier in this chapter, I mentioned how airside systems could use variable speed/frequency drives to vary the speed of the fan.

As the fans speed is varied, the amount of airflow will increase or decrease. It's important to note that the increase and decrease of airflow is not linear. This means that the amount of airflow coming from a fan at 75% speed may be substantially more than the amount of airflow coming from a fan at 50% speed.

To determine the amount of airflow, that will be produced you should refer to what is called a fan curve. A **fan curve** shows the amount of CFM that is moved by the fan based on the RPM of the fan.

It is important to note that fans have different fan curves based on how they were designed.

Coils

Coils exist to transfer heat to and from the air. Hot or chilled water will flow from the waterside systems to coils that are installed inside the air side systems. The valves that are mounted in the piping before the coils will modulate the amount of flow that reaches the coil. As the air moves through the coil, it will either gain heat or transfer heat depending on what type of water is running through the coil.

It's important to know that there are two main control methods for using coils. Those methods are the two-pipe and four-pipe system.

In a **two-pipe system,** the hot water and chilled water use the same piping. To switch between heating and cooling, the facility will have a changeover setpoint. Typically, this set point is called a summer\ winter changeover. This changeover will determine whether chilled or hot water is being supplied to the coils. While this does reduce the amount of piping that the system needs, it also means that you cannot have heating and cooling running at the same time.

This is okay for climates that do not have big swings in temperature, but this control method may create challenges for climates that do have big swings in temperature.

A **four-pipe system** provides dedicated piping for both hot and chilled water. This means that an air side system can utilize both heating and cooling.

Refrigeration Cycle

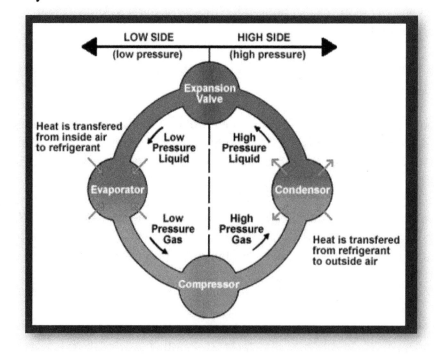

One of the key concepts I want you to take away from this chapter is the concept of the refrigeration cycle. The refrigeration cycle is used in so many HVAC systems.

I also find that understanding the refrigeration cycle will help you to understand how your chillers and rooftop units' work.

As I mentioned earlier in the chapter refrigerant changes state at different temperatures than water.

Because of this, you can take heat from air, water, etc. and transfer it to the refrigerant. Once this is done, you can also transfer the heat from the refrigerant to a different mass of air/water.

It is important to know how the refrigeration cycle works in case you need to work with a technician or you need to understand why a designer is implementing a specific design.

The refrigeration cycle consists of two sides, a low side, and a high side. When you hear folks talk about the low side of the refrigeration cycle, they are often referring to the side of the cycle that absorbs heat. The low side is also known as the evaporative side.

When you hear folks mention the high side of the refrigeration cycle they are often referring to the side of the cycle that exhausts heat. The high side is also known as the condenser side.

A refrigeration system consists of 4 main pieces:

- Expansion valve
- Evaporator coil
- Compressor
- Condenser coil

There are four steps in the refrigeration cycle.

Step 1:

Starting on the high side, the expansion valve restricts the flow of refrigerant. This causes the refrigerant to become a low-pressure liquid. This low-pressure liquid now can enter the evaporator coil.

Step 2:

The low-pressure refrigerant enters the evaporator coil which uses the principle of evaporation to transfer heat from the air/water to the refrigerant. This process causes a state change which converts the liquid refrigerant into a low-pressure gas.

Step 3:

This gas goes into a compressor which now compresses the gas into a high-pressure gas. This gas then goes on to the condenser coil.

Step 4:

The gas enters the condenser coil which will use the concept of condensation to convert the gas back to a liquid this process will transfer the heat from the liquid to the outside air or condenser water. This high-pressure liquid now travels back to the expansion valve.

That my friends is the refrigeration cycle. Not very complex eh?

I mean sure the theory stuff is complex, but the process is pretty cut and dry.

Now when you hear the term compressor or evaporator coil, you will understand what these devices do.

Psychrometrics

Psychrometrics is quite possibly one of the most misunderstood aspects of building automation systems. I remember when I was going through HVAC boot camp and the instructor pulled out a psychrometric chart. You would have thought that he had just asked the class to perform brain surgery.

Now, if some of you are wondering what Psychrometrics is. Don't worry. You're not alone.

Psychrometrics is the study of the level moisture in the air at different environmental conditions. During your building automation career, you may or may not have heard of a psychometric chart.

The **psychrometric chart** is used to help map out the effects of heating and cooling on the moisture level of air. Don't worry I'm not going to teach you how to read a psychrometric chart in this book.

Now, why is this important to you?

This is important because human comfort is largely tied to the amount of moisture that exists in the air. Like I said earlier in the chapter, I used to live in Houston. During the summers, Houston would get ridiculously humid.

Now there was this one time I had to travel to Dallas. For those of you who aren't familiar with the great state of Texas, Dallas is about 240 miles north of Houston. When I left Houston, it was roughly 90° and 80% relative humidity (RH). Four hours later when I arrived in Dallas, it was now 98° and 40% humidity.

Surprisingly, Dallas felt rather cool. The reason behind this was due to the lack of moisture in the air allowed my body to take advantage of a process called evaporative cooling. **Evaporative cooling** is the transfer of latent heat into the air.

You may be wondering what latent heat is. **Latent heat** is heat that is released during a state change.

If you've ever brought cup full of hot tea or coffee outside during the winter, you've probably seen this in action. As the heat from the liquid begins to transfer into to the air, in the form of moisture. This moisture then condenses into a gas above the cup.

You've probably also seen this if you've ever driven by a cooling tower on a cold morning.

Ok, I feel like I need to back up a bit because in those few short paragraphs I described several key concepts that I need to unpack.

Begin with the comfort zone

According to ASHRAE 55-2010, there is a certain range of air and relative humidity that the average person finds comfortable. This range is called the **comfort zone**. The comfort zone is between 70° to 76° dry bulb temperature and 45% - 65% relative humidity.

Now, I am going to describe some terms in this section that may be unfamiliar to you, if they are, don't worry, by the end of this chapter, they won't be. These terms are:

- Dew Point
- Dry Bulb
- Relative Humidity (RH)

A little story

Rather than boring you to tears as I define each of these terms, I'm going to tell you a story to illustrate these points.

Imagine you are the facility director for a large campus. You are driving to work with the windows down enjoying a beautiful morning. The air outside is a perfect 72° (this would be dry-bulb temperature), and the relative humidity is 60% (this means that the air is 60% full of moisture, relative to the amount of moisture that the air can hold at 72°).

However, when you enter the building, you notice that the air inside the building is 80° and 80% relative humidity. You realize you need to cool the air down and remove some of the humidity before folks show up for work.

Now could you just go and use that cool outdoor air to cool the building down?

Maybe.

As the air gets hotter the amount of moisture, it can hold increases. So, 80% relative humidity (RH) at 80° is a lot more moisture than 80% RH at 72°.

So let's say you start to mix the 72° air with the 80° air.

Do you know what the result would be?

You could increase the amount of moisture in the air. As you can imagine increasing the humidity level of the building is a problem.

To remove moisture from the air, you need to take advantage of something called condensation. **Condensation** is when the moisture in the air condenses to a liquid. This typically occurs when the air temperature reaches the dew point temperature.

So, what is the dew point?

The **dew point** is the temperature where the air is fully saturated. When I say the air is saturated what I mean is that the air cannot hold any more moisture. As the air gets colder, the amount of moisture it can hold is reduced.

This is where mechanical cooling comes in. You can use your air conditioning to cool the air to the dew point temperature which will cause the moisture to be removed from the air.

By the way, this is why you see so many HVAC airside systems setup with a set point of 55°.

Chapter 1 Quick Summary

Awesomeness, so we're through our first chapter together, schweet! This chapter gave you a quick primer on what you need to know about HVAC systems. Now, I'm going to take a quick moment and recap what I covered in this chapter.

First, HVAC systems include airside and waterside systems. Airside systems, include air handlers and terminal boxes. These systems move the air around using fans. Waterside systems exist to heat up or cool down the water that these airside systems use.

There're a couple of ways that airside and waterside systems can work. Together we discussed how air pressurization, balancing, and water distribution loops work. Sure there's more to this topic than that, but, I'm trying to prepare you for the BAS world, and HVAC is just another piece of the puzzle.

Finally, I dug into some complex topics including the refrigeration cycle and psychrometrics. My hope is that now you have a solid understanding of how those two topics work.

Alrighty, are you ready to take the BS out of BAS?

Oh, wait, that's not right, that just leaves us with an A.

Hmm...

Moving on...

In the next chapter, I'm going to take you from countdown to liftoff. Get ready to ignite your learning engines because your BAS knowledge is about to go through the roof!

CHAPTER 2

THIS IS YOUR BUILDING AUTOMATION SYSTEM

What's in this chapter?

This chapter serves as a primer for the rest of the book. In this chapter, I will shore up any gaps in your understanding of Building Automation Systems (BAS).

This chapter will cover:

- A brief history of building control systems
- The parts of a building automation system
- Key functions of a building automation system
- Common protocols that building automation systems use
- Control logic fundamentals
- Common control sequences

The business value of a BAS

Let me ask you a few questions.

- Can a BAS make help you avoid losing money?
- Is there a business value associated with a BAS?
- Does the way you operate a BAS impact your business outcomes?

I argue that the answer to all three of those questions is a resounding yes! And this is the point I want you to capture. As you read through this chapter, I want you to think of the business impacts associated with each topic.

Here's why.

If a building can't be used due to the failure of the air conditioning, lighting or some other system, then the owners are going to lose money. However, I've got some good news. My experience has shown me that most BAS failures can be addressed through 3 things:

- Design
- Training
- Service

As you move through this book, I am going to teach you the fundamental knowledge you need to oversee the design, installation, and servicing of a building automation system.

The journey of a thousand miles begins with a single step

To appreciate where the controls industry is it helps to know where the industry came from. The controls industry started a long time ago with wax thermostats and all sorts of other contraptions. That information, however, is useless to you, and this is a useful book!

Therefore, I will start by walking you through the building control systems that you may encounter on your journey. My intent is not to make you an expert on each of these systems. Rather I want to give you a familiarity with these control systems and how they apply to your current situation.

A brief history of building control systems

There once was a guy who made a thermostat. Folks liked it. So he made more.

The End.

Ah, if only life were that simple.

I'd love to tell you, that you only need to understand the current flavor of control systems and call it a day.

However, I'd be leaving you woefully unprepared for the rat's nest of systems that you will encounter on your journey.

That's why I'm starting this chapter off by covering the family-tree of control systems. In this section, I will be taking you from pneumatics all the way up to the current generation of direct digital controls (DDC).

As I move through each of these systems, I want you to ask yourself these two questions. (Don't worry, I will answer these questions for you later in the book):

- How could I replace this?
- How could I maintain this?

Don't skip this section! You'll thank me later when you encounter these systems in your day-to-day happenings!

Now, before I move into the topic of control systems, I want to address a thought some of you may have.

What is the difference between a control system and a building automation system?

The difference between a control system and BAS is, a control system describes the physical aspects of a system, and a BAS describes the control and interface aspects of a system.

To envision this, picture a network.

A network can have several different physical layouts. For example, a network could use a bus or a star pattern. This physical layout would be the "control system."

Now, to control the messaging flow, the network needs to connect back to a device. The most common device in networks is a switch. A switch will control all of the messaging and allow the users to change network settings. In this example, the switch would be the "control and interface" point.

This would be similar to a BAS.

Ok, I know some of you may not know what a switch is. I purposely used an example that I knew some of you wouldn't be familiar with.

As you go through this book, you will notice that I use a lot of analogies. I want you to know I am purposely doing this to get you used to the terminology.

Alright, back to control systems.

The eras of control systems that I will cover in this section are:

- Pneumatics
- Electromechanical
- Analog
- Digital
- Direct Digital Controls

Pneumatics

I still remember replacing my first pneumatic valve. It sucked, or maybe I should say it blew?

Here I was rebuilding a valve with a $2,000 repair kit when I could buy a perfectly good electrically actuated valve for $400. This simply did not make sense to me.

That my friends is the world of pneumatics. They leak air, they get water in their air lines, they cost a lot of money to repair, and finding folks to work on them is quite difficult.

So, why are pneumatics still around?

Two words, **retrofit cost**.

The cost of retrofitting pneumatics is quite high. Replacing a pneumatic system with a direct digital control (DDC) system normally requires a completely new control system.

A normal pneumatic to DDC upgrade will require new actuators, controls, wiring, programming, and graphics.

When budgets are tight, a full retrofit is often not cost effective.

How does it work?

So how do pneumatics work?

Have you ever tried to add water to a cup of water after it was full?

The water has nowhere to go, so it makes a pretty big mess.

Well, imagine if that cup was sealed tight and it had a hose coming out of it. If you tried to put more water in the cup, the water would go out the hose. This is due to the weight and volume of the water.

Well, guess what?

Air has volume and weight as well.

While air is not nearly as dense as water, it still has mass. Compressed air, which is what pneumatic devices use, can exert an amazing amount of force. It is this force that moves dampers and valves.

A pneumatic system compresses air using a device called a compressor. A compressor is an electronic device that compresses air, trust me that's all you need to know.

This compressed air will flow through the pneumatic tubing to a piece of HVAC equipment. It is here that a sensing element allows air to enter or bypass the valve or actuator.

Using an overlay method for a partial retrofit

Now, in some cases, folks will use an overlay method to replace pneumatics with DDC. An **overlay method** allows the system owner to gradually replace the pneumatic system, one device at a time.

There are ten steps to executing an overlay method. These steps are:

- **Step 1:** Purchase the DDC system
- **Step 2:** Setup the DDC system and install the field controllers
- **Step 3:** Identify the control devices and sensing elements
- **Step 4:** Work with an electrician to run wire to the sensors
- **Step 5:** Put the pneumatic system in hand
- **Step 6:** Install E2P transducers to replace pneumatic systems
- **Step 7:** Add new sensors and control devices where needed
- **Step 8:** Install the field controllers and wire up all E2P, sensors, and control devices where needed
- **Step 9:** Test the new DDC system
- **Step 10:** Release the systems from hand

As I said, there are ten steps to performing an upgrade. I briefly describe this process below.

The first step in the overlay method is for the system owner to purchase a DDC system.

Next, the DDC system is setup, and a local field controller is put in place.

A controls technician will identify all of the control devices (valves, actuators, inlet vanes) and the sensing elements (temperature, pressure, airflow). Once this is done a technician or electrician will run a wire to each of the identified devices.

Next, the pneumatic control devices on the system are inventoried, and a device called an electrical-to-pneumatic (E2P) transducer is installed next to each pneumatic control device

The E2P transducer works by taking an electrical signal from the DDC field controller and driving a diaphragm or valve to regulate the compressed air that is fed to the device.

After the E2P is installed, new sensors will be added. These sensors, along with the E2P devices, will be connected to the wiring.

After all of this is done the HVAC system and its new controls will be tested, balanced, and put back into operation.

Electromechanical - not quite vacuum tubes

Do you have a garage door at your house?

Does your garage door have an automatic opener?

If you answered yes to both questions then you my friend are the proud owner of an electro-mechanical system.

An **electromechanical system** is a system that uses electricity to make mechanical systems move. In the case of your garage door, you press a button on your garage switch and a relay inside that switch will allow electricity to flow to your garage door's motor.

These kind of systems are still surprisingly common. If you've been inside a warehouse, then you've probably seen wall switches being used to control temperature. The wall switch allows the owner to set a temperature setting for their space. If the temperature setting is exceeded, a device, typically a heater or an exhaust fan, will kick on to control the temperature.

As you can imagine these kind of systems are limited in the control capabilities that they can provide you. The benefit of these systems is that they are dirt cheap. For simple spaces like a warehouse or a broom closet, an electromechanical device usually does the trick!

Monitoring electromechanical controls

One area that you will need to address when dealing with electromechanical controls is the area of monitoring.

The stand-alone nature of electromechanical controls enables these controls to be cheap, but it also prohibits them from communicating their values back to the end-user. Since these controls are still around, many manufacturers are shipping their electromechanical controls with 0 to 10 volt direct current (VDC) outputs to supply a feedback signal to modern control systems.

Key point:

You need to consider monitoring when you design your control standards or evaluate controls proposals. If monitoring these controls is important to you, then you should make sure you either include this requirement in your standards or ask for it in your bid solicitations.

Analog controls

An analog building automation system is hard to find. The funny thing is when I first wrote this section I couldn't even think of an example of an analog system.

Analog systems work by sending variable signals across a wire. A good example of this would be to think of an electronic valve that is directly controlled by a temperature sensor. The sensor provides an analog signal that the valve interprets and uses to control the valve.

In my experience with hundreds of control systems, the closest thing I've seen to true analog controls is a potentiometer. A **potentiometer** is a manual or automatic device that, adjusts the resistance across a wire. This adjustment is based on either a manual or automatic input.

When you do encounter these systems, it's usually best to replace them with modern controls and sensors. There is no cost effective way to integrate analog controls.

Digital controls

Digital systems were a step in the right direction.

Digital systems produce a signal that is either a zero (off) or a one (on). With these signals, you are able to talk to computers and other processors directly.

The adoption of digital controls was a big deal because the controller normalized inputs and outputs into digital signals that could be processed by the actual controllers.

There is a blurry line between digital controls and direct digital controls (DDC). An easy way to remember the differences between the two is to remember that digital controls do not allow distributed control of devices across a network.

There are many, many digital systems still installed in buildings right now. Between pneumatics and digital controls, you are very likely to encounter a legacy system in your career. This then brings us to the next logical question.

How can you integrate these controllers into your current BAS?

There is only one path when dealing with most digital systems, and that is to replace the controller. The reason behind this is because the majority of digital systems do not have any onboard communications capabilities. Because of this, even if you did want to connect this device to a modern network you couldn't.

However, as with anything in the BAS world, there are exceptions to this.

Some digital systems do have the ability to connect to a gateway that the manufacturer has developed as a "bridge" between the old and new controls system. This is why I highly recommend talking to the manufacturer before you decide just to rip everything out.

Direct Digital Controls

Direct Digital Controls (DDC) take the concept of digital control and move it to a distributed model. Now, I've used the term distributed model several times now, so what exactly does that mean?

The concept of DDC is that your field controllers can execute their control logic (sequences, commands, etc.) locally and then report all of that information back to a central supervisory device.

Before DDC, the controls solutions we had required either manual control or elaborate setups to totalize the outputs from field controllers into something that would make sense to the average user.

With DDC, the controllers themselves now have the capability to run programs and communicate with one another. This was a huge shift in the market.

In the next section, I will describe the common functions of a BAS and then I will explore the architecture of a traditional DDC System.

Common functions of a BAS

What does a building automation system do?

What should it do?

I often find myself helping business owners with properly identifying the functions of their current building automation system. I then work with the business owner to help them formulate a vision of what they want their building automation system to do in the future.

At a minimum a building automation system should be capable of:

- Alarming
- Command and Control
- Monitoring
- Trending

Alarming

In my experience, alarming is the most mismanaged capability of a BAS. If I had a dollar for the number of Fortune 100 companies that I walked into who had 50,000 + unacknowledged alarms, I'd have almost a hundred dollars.

The reason alarms get so out of control has to do with how alarms are set up. In most projects, alarms are about as well thought out as my dinner planning skills. The reality is on any given project an alarm specification is often written like this,

"The BAS will alarm when space temperature exceeds space set point by 2° F (adj.)"

Do you see the problem with that statement?

It's ok if you didn't because at first, I didn't either. The wording above caused my customers and I a ton of headaches.

Think about the specification language I just showed you for a second. Does that wording tell you what happens when the space isn't occupied?

Does the alarm trigger then?

Is the alarm still set to 2° adjustable?

Don't worry, later in this book, I will teach you how to create standards so clear, my great grandma Nana could configure them.

In addition to poor alarming thresholds, there also tends to be a problem with poor alerting.

Who does the alarm go to?

What happens when the alarm isn't acknowledged?

These are things that I have yet to see written into a design specification.

Now before it comes across as me being anti-consulting engineer, I'd like to speak in the consulting engineers defense. How can a consulting engineer be expected to design a standard that meets this requirement, when often time's the owners have not considered these issues?

I always tell my customers that the design phase is your time to ask for your wish list. I can tell you from experience that it takes the same amount of time to set up an alerting matrix as it does to just set up an alarm.

I hope that you leave this section with the realization that if you aren't getting your alarms setup right, you're just creating extra work for yourself down the road.

Command and control

No, I'm not talking about Command and Conquer, kudos to any of you who got the reference to the 1990's video game that consumed a completely ridiculous amount of my life.

Command and control is the capability that allows you to…

Drum roll, please, command and control your building automation system.

As you can imagine, this capability is so critical it goes without saying that you need this capability.

So how can I define the capabilities of command and control?

I like to lump command and control into three buckets:

1. Ability to create user roles
2. Ability to override points
3. Ability to schedule points

Create user roles

The ability to create user roles is critical.

User roles allow you to give or limit access for your BAS users to command, view and modify your BAS points. Later, when I get into the user interface portion of this book, I will describe how to setup the building automation system to provide separate user interfaces. This will dramatically reduce the learning curve for a BAS.

Override points

No matter how good a system is at some point, the user will need to override a point. This feature is especially useful for temporary overrides and should be one capability that is specified in a company's building automation standards. After all, many a good control system have been wrecked by over-zealous users who forgot about the overrides they put in place.

Schedule points

Scheduling is the final command and control feature. You will want to have a robust scheduling system as I'm sure you don't wake up every day planning to go into your business to turn on and off building automation systems.

Monitoring

There's an old saying that, what you measure gets done. I like to say that in the world of BAS, it's what you trend and alarm that gets fixed. I've seen control systems go years with leaky valves, inoperable dampers, and clogged filters.

The thing is, each one of these systems is costing the building operator thousands in maintenance and energy costs each year. Many control systems are installed without a reliable, structured way to detect problems, notify personnel of the problems, and troubleshoot the problems.

The first step in solving this problem is being able to monitor your BAS.

Graphics, also known as **Graphical User Interfaces (GUI's)**, provide a way for the operator to visualize how their systems are performing. There are three types of graphics that you should know about:

- Floor-plan graphics
- System graphics
- Room graphics

Floor-plan graphics

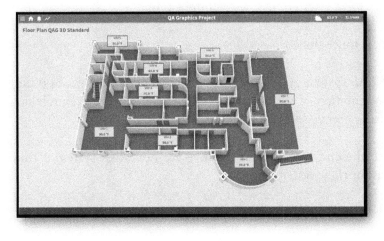

In the world of BAS systems, floor-plan graphics are quite common.

The purpose of the **floor-plan graphic** is to show the conditions of your building in relation to the actual floor-plan. In some cases, these graphics will provide a "heat-map" as an additional layer on the floor-plan. A **heat-map** utilizes different colors to help the user visualize the difference in temperature from a set value, usually 72°.

A heat-map provides a quick way for the end-user to evaluate the health of their building at a macro level.

System graphics

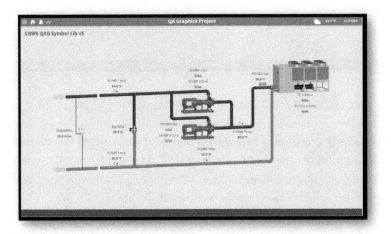

A **system graphic** is a graphic of a system, but what exactly is a system?

A **system** is a collection of devices or, in some cases, a single device that provides a specific function. If that wasn't clear, don't worry I'm going to break that down for you.

Take a moment to think about a car engine. A car's engine has cylinders, gears, belts, etc. All of these parts come together to form a system that we call the engine.

System graphics are quite similar.

An example of a BAS system graphic would be a chiller plant or rooftop unit graphic. This graphic would contain a representation of the physical layout of the system, the individual devices that make up the system and the data points associated with this system.

From this graphic, the end-user would be able to see how their system is running and would be able to adjust the setpoints for their system.

Room Graphics

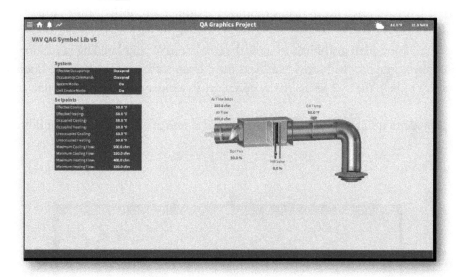

The final graphic type that users will commonly encounter is the room, or space, graphic. •

The room graphic usually contains a representation of the terminal unit, often a Variable Air Volume (VAV) box that is supplying conditioned air to the room. This graphic usually includes the image of a thermostat which displays the current conditions within the space. This brings up a key concept with graphics, and that is the concept of templates, also known as aliased graphics.

As you would imagine developing graphics for 2,000 rooms that have, identical VAV box setups would be mind-numbingly repetitive.

That's why graphics developers created the concept of templates to avoid this. In a **template graphic,** the points are "aliased" or genericized with a wildcard character (*). The graphic is then linked to individual field controllers, and the wild-card is replaced with the controllers Fully Qualified Reference (FQR), which is a fancy way of saying the controller's name.

If all of this seems a bit overwhelming, don't worry. Later in this book, I will provide you a framework for creating standards around your graphics.

Trending

I want to take a second and go a step further on the concept of monitoring by discussing trending. **Trending** is the art of logging specific data points, called **trends**, and storing them in a database so that you can evaluate and report on them at a later date.

The reason I called trending an art is that trending requires experience to know what points to setup trends on and how often to trend these points for.

If a proper trending strategy is not developed, a BAS can become bogged down by useless data that consumes storage space without providing any value.

Trends tend to fall into two categories. These categories are time-based trends and change-of-value based trends.

Time based trends

Time-based trends are trends that are setup to record a specific data point based on a time interval. Once this time interval is reached a "sample" of the data point is captured. This usually involves capturing the "present value" of the data point and storing it locally on the supervisory device in a queue.

In the BAS world, a **queue** is a storage location for data that is local to a device. Once this queue reaches a specific threshold, the data is sent in a **batch** or a collection of data samples. This data is then transferred to the database where it can be recalled at a later time.

Change of value trends

Change of value trends are exactly what they sound like. A **change of value (CoV)** trend is different than a time interval trend. Using CoV the BAS will monitor a data point and record a sample whenever the value of the data point changes by a set value (hence the name change of value). Once these trends are logged into the queue, they will follow the same process as time-based trends in how they are sent to the repository.

Trending is an art, and you need to put thought into how trends are setup. You should be asking yourself questions like, what should my data collection frequency be and how can my trends be viewed.

Parts of a BAS

Modern Building Automation Systems tend to be designed using either a three-tier or four-tier model.

In this section, I will describe what those tiers are and what they do. In addition to those tiers, I will also teach you what the different types of field controllers are. Finally, I will close this section with an overview of the most commonly encountered control and sensing devices.

After you read this section, you will understand what the different parts and pieces of a BAS are.

Three-tier BAS model

The **three-tier model** is a model that is commonly found in smaller stand-alone BAS systems. I have usually found this model being utilized in small commercial office buildings and school districts.

In the three-tier model the supervisory device, covered in the next session, acts as the "server" by providing access to graphics, trends, alarms, and scheduling.

Pros

The pros to this design are reduced cost and complexity. Since there is no server the cost of installing this model is usually less expensive than the four-tier model.

In addition to the lower cost, there is no server to maintain. This makes the cost and complexity of supporting this solution lower as well.

Cons

The cons of this design are that the supervisory devices often have limited alarm and trend capacity. This means that the supervisory devices can only store trends for a limited amount of time, usually a few weeks. Also, the ability of this model to support multi-site deployments is greatly limited.

Four-tier BAS model

The **four-tier model** introduces a dedicated server into the BAS architecture. This server provides several features. I am going to focus on three of the functions that a server provides. These are:

- Data storage
- Reporting
- Multi-site management

Data storage

As I mentioned earlier, BAS systems log events and changes using trends and alarms. The trend and alarm data require storage.

One of the main purposes of a server is to oversee the transfer of that data into a database. I will cover databases in greater detail in Chapter 4 but suffice to say, **databases** take data and organize it into tables. This allows the data to be retrieved at a later date.

Once the server has gathered the data from the field controllers, it is responsible for overseeing the processing and normalization of the data. This must be done before sending the data to the database.

While you do not need to be a Database Architect to run a BAS you should understand how the server collects data from the field devices and then sends this data to the database.

If someone wants to look at the alarm or trend data later on the server will connect to the database and request the data. Reporting is one of the major tasks that is performed on a daily basis at larger BAS installations. Before I discuss reporting in the next section, I wanted to point out that reporting requires data management and data storage.

Reporting

Good ole' reporting.

Do you want to know which rooms in your building are more than 4° from set point?

Reporting can do that.

Do you want to know the total energy consumption of a particular floor?

Reporting can do that as well.

The capabilities of a BAS server come in handy when you want to analyze and report on historical data. Most BAS systems support trend analysis and data reporting. While there are some BAS systems that don't perform these capabilities, I would argue that those systems aren't a BAS.

There is a part of me that feels like I should write more on this subject, but honestly, reporting is pretty simple. You can setup a report by defining your data sample (the data you want to report on), the period of the time you want to report on and the format you want the data returned in.

Maybe at a later date, I will write a book dedicated to trending, reporting and data analysis.

Multi-site management

In Chapter 4 I will cover the concept of networks in detail but for right now all you need to know is that multiple sites can be connected to a network. However, I should probably define what a site is before I dive any deeper into the concept of multi-site management.

A **site** is a single physical location that may contain multiple supervisory devices. Multiple sites can be connected by what is often called, ready for it…

A site.

Yes, if you are totally confused by that terminology you're not alone. I still catch myself having to remember to clarify do folks mean site or site?

Now some BAS call their sites stations or supervisors but by and large the most common term I run into is site.

Ok, with that brief segue done, let's get back to the topic of multi-site management.

When you move beyond a single physical location, you will most likely want the capability to manage multiple sites on a single graphical user interface. To do this, you need to have a central site that manages the connections between the server and your local sites.

This is often done using a concept called a heartbeat or a keep-alive. A **heartbeat** is a message that is sent from the local sites to the server and back. If this heartbeat message is not received in a certain amount of time the site will indicate that it is "offline." In addition to the heartbeat concept, a server can also send out a time synchronization message.

The purpose of a time synchronization message is to ensure that all of the sites are basing their messaging and data forwarding based on the server's time. Without this, you could have local sites that were sending data that was not "in sync" with the server. This would cause major problems when it came to reporting, alarming, trending or centralized control logic. This is because any data that was from the local sites would be out of sync with the central server. This can cause all sorts of nastiness.

Time-sync related issues are one of the most common issues I encounter when I am helping folks work on their multi-site installations.

Building automation system hardware

Supervisory device

A **supervisory device** is usually a full-blown computing device that provides graphics, scheduling, alarming, and trending. The supervisory device can also provide remote connectivity and remote alarming.

Typically, a supervisory device will contain a traditional operating system like Linux or Windows. Supervisory devices are capable of providing stand-alone control for smaller systems. It's quite common to see a single supervisory device controlling a small office building or school.

Field controller

Field controllers are specialty devices that use inputs and outputs to monitor and control devices. Often these controllers will be capable of supporting complex programming. Some controllers are capable of operating in a stand-alone mode if the supervisory device fails.

A field controller is often connected to a supervisory device via a field communication bus. The most common field communication bus right now is BACnet MS/TP (Master/Slave Token Passing). The field controller(s) use the communication bus to communicate to other field controllers and supervisory devices.

As I mentioned earlier, a field controller will have inputs and outputs.

Inputs are how the controller senses the status of the control devices (status switches and sensors for elements like temperature, pressure, and humidity). Usually, these are located on the left side or top of the controller.

Outputs are used for controlling devices (actuators, relays, etc.). Outputs are often located on the right side or bottom of the controller.

The field controller often has a processor inside it that executes code based on the sensed values of its inputs. Sometimes a field controller can act based on inputs from other devices. These are called **network variables**. A field controller can use network variables to receive inputs from other controllers and to send outputs to other controllers.

Field controllers are normally powered by 24 volts alternating current (VAC) or 24 volts direct current (VDC). You will occasionally see controllers powered by 120 VAC but in my experience these are rare. One other method you may see used to power controllers is called Power over Ethernet (**PoE**). I will cover this method in greater detail later on in the book.

There are three main types of field controllers that you should be focused on these are:

- Free-programmable controllers (FPC's)
- Application specific controllers (ASC's)
- VAV modular actuator controllers (VMAC's)

Free-programmable controllers (FPC)

A **free programmable controller** is a controller that is freely programmable.

Ok, great, what does that mean?

In the past, field controllers were developed to perform a specific application. These controllers were called application specific controllers, and I will cover them in greater detail in the next section.

As the abilities of BAS controllers increased, the types of sequences that the controllers could implement also increased. This resulted in complex control sequences that the standard application specific controller could not address. The free programmable controller was developed to solve this challenge. This controller allows the BAS technician to create custom logic based on the needs of his or her specific sequence.

Funny enough, as I write this, the free programmable controller is caught up in one of the larger debates in the BAS world. You see, folks don't want to be locked to a specific controls vendor. However, most of the field controllers can only be programmed using a vendor's tools.

Now, despite whatever BS you've heard from so called "experts" this isn't because the controls companies want to lock their customers. Trust me there is plenty of vendor locking but not in the actual field controllers.

This has to do with the way that embedded controllers are developed and the source code that is used for a lot of the controllers.

Even if the whole industry switched to home automation boards tomorrow, we would still have the problem of transferring our code base to other controllers.

To have a common programming software all of the controls companies would have to agree to use a single field controller, developed on a single code base. Which would, in turn, lead to customers being locked to only one field controller.

I want you to realize that no matter how much of this "open controller" concept crap you hear floating around on the Internet, to have a common framework for a field controller everyone will have to agree to buy the same field controller hardware and program it the same way.

This means you will still be locked to a single controller.

Remember to be cautious when folks promise shiny things…

Application specific controllers (ASC)

The application specific controller is the opposite of the FPC. ASC's are often made to serve a single purpose. They may allow some customization through a programming menu, but they cannot be freely programmed.

The inability to freely program ASC's is not a bad thing. There are often specific types of equipment that are well suited for ASC's. An example of an ASC would be a dedicated rooftop controller.

VAV modular actuator controllers (VMAC)

The VAV modular actuator controller or VMAC is a hybrid controller. The VMAC is physically designed for VAV boxes. From that perspective, the VMAC is an application specific controller. However, the VMAC is also an FPC because it allows the technician to freely program it. Physically, VMAC's come with a damper actuator built into the controller and are designed to mount on terminal units.

Sensor/Actuator

While the sensor/actuator layer may seem fairly self-explanatory, it contains quite a bit of information. This layer allows for monitoring and control of things like temperature sensors, actuators, relays, etc. In

Chapter 1 I discussed several HVAC components. In this section, I will cover the sensors and control devices that control those components.

The Sensor/Actuator layer consists of two groups of devices. These groups are:

- Control devices
- Sensors

Control devices

Control devices are normally associated with outputs.

As I mentioned earlier, a field controller reads an input value and then tells an output what to do. This is very similar to your home thermostat which senses the temperature in your house and then tells your home air conditioning unit whether to turn on the heat or the cooling.

I've divided outputs into two categories. There are more categories, but these are the main three you will encounter. These categories are:

- Actuators
- Relays

Actuators

Actuators are mechanical devices that use a motor to rotate a device. Typically, an actuator is used to control a valve or damper. An actuator works by receiving a signal from the field controller's output. The actuator will then use that signal to drive the actuators motor. The motor will rotate what is called a stem or shaft to a specific position. The stem or shaft is a rod attached connected to a mechanical device, like a valve or damper.

There are three types of actuators that I will cover. These types are:

- On/off
- Floating
- Proportional

Each of these actuator types will have one of two failure options:

- Spring-return (SR)
- Non-spring Return (NSR)

Spring return versus non-spring return

Every actuator is designed with a fail-to-position.

A **fail-to-position** is where the actuator will fail to when its power is removed. Some actuators are designed to fail open. To fail open, the actuator must have a way to return to that open position. This is where return spring comes into play. When an actuator with a return spring fails, the spring will return the actuator to its previous location.

On/off Control

On/off control is just like it sounds.

With **on/off control**, 24 to 277 VAC is provided to the actuator to drive it open. In this form of control, the power is sourced from a transformer through a relay. The BAS field controller will energize the relay to allow the source power to pass-through to the actuator.

Don't worry if this is confusing. I will cover relays in just a second.

Floating control

Floating control allows the actuator to float between two positions by having two motor solenoids that drive the actuator. Also known as three-wire control, floating control has a common wire and two power wires one for clockwise rotation and one for counter-clockwise rotation.

The field controller will typically use what is called a sequencer block in its programming to "sequence" the open or close output to trigger.

Proportional control

Proportional control uses a separate control signal to tell the actuator what position to dive to. A proportional actuator has a separate control input and power input.

The actuator will get 24 to 277 VAC from a nearby power source. The field controller will typically provide a 0 to 5 VDC or 0 to 10 VDC signal to the actuator. This signal is often proportional to the % that the controller wants the actuator to drive, hence the name proportional control.

For example, a controller that wanted an actuator to drive to 50% open would provide 5 Volts DC if the controller was using a 0 to 10 VDC signal.

Relays

As I mentioned earlier, a controller can supply power to devices. However, controllers are only capable of running so much amperage through their outputs before they "let their smoke out" which is an industry phrase that means the output melted.

How can a controller supply 24 to 277 volts to the devices they need to control without cooking itself?

To solve this problem, technicians will use a device called a relay. In this section, I cover the two relay types that I have used throughout my career. If you are an electrical engineer by trade, then this section may seem quite basic, but it is a key concept that anyone touching a BAS should understand.

The two relay types I have worked with are fusible and non-fusible relays.

Quick note

For those of who don't know what a fuse is, a **fuse** is a component with electrical connections inside it that are designed to break, usually by melting, when a certain level of amperage is present.

Amperage is essentially heat.

When electricity travels across a set of wires, it encounters resistance. As the electricity increases and the resistance stays constant the amount of amperage increases. You can experience this by dragging your foot across a rough surface.

Do you feel the resistance?

If you rub your foot or hand fast enough, it will start to get hot from the resistance.

Fusible relays

You will see fusible relays when you are working with high amperage devices like fan motors and large dampers. The purpose of a fusible relay is to allow the fuse to be destroyed before the actual control components are damaged.

Non-fusible relays

Non-fusible relays are the most common relay you will encounter. A non-fusible relay is simply a relay without a fuse.

How do relays work?

Relays operate based on the concept of a pole and a throw.

A **throw** is a circuit, and a **pole** is the motor to drive the gates within that circuit. When power is applied to the pole, it produces an electrical field that will move the throw.

Often you will hear the term single-pole-double-throw or SPDT. This means that the relay has a single pole that can have power applied to it and two throws that can be operated.

An SPDT relay will often have a throw that is normally-open (NO) and a throw that is normally-closed (NC). When power is applied to the pole, a normally-open throw will close, and a normally-closed throw will open.

Sensors

Honestly, I hesitated to even write this section. After all, there are so many different sensor types, how could I even do justice to this topic?

Therefore, I will preface this section by saying that these are not all of the sensors you will encounter. I took the approach of looking at the common types of sensors you will encounter. Based on my experience the most common sensor types you will encounter are:

- Resistive sensors
- Powered sensors
- Circuit or switch sensors

Resistive sensors

Resistive sensors are sensors that utilize resistance to determine a value. The most common type of resistive sensor is the thermistor or temperature sensor.

Resistive sensors work when a control voltage, usually 5 Volts DC is applied across a wire, as the voltage returns the resistance applied against the voltage is detected, and this resistance is then converted to a temperature reading.

With resistive sensors, the important thing to note is that there are two main types of sensors, Positive Temperature Coefficient (PTC) and Negative Temperature Coefficient (NTC).

A **PTC** sensor will increase its resistance when the temperature it measures increases.

An **NTC** sensor operates by increasing its resistance when there is a decrease in temperature.

Understanding whether your sensor is PTC or NTC is important when you are troubleshooting sensors. If you are testing your sensor, you need to know what an increase or decrease in resistance represents.

I can't tell you how many times I got called out to troubleshoot "bad" sensors, just to find out that the technician had configured the sensor as a PTC sensor when it was actually an NTC sensor.

The final note on resistive sensors is that of calibration.

While calibration is important for all sensors, it is very important for resistive sensors. Over time resistive sensors can drift which can cause inaccuracies of up to 1% of the sensor range (or 3°-4°). This means that you could be overdriving a cooling valve on an air handler to cool air that is actually at set point.

For a typical office building, this can equate to over $1,400/yr. in excess conditioning costs over the course of a year. The scary thing is, this is just one sensor.

Because of this, I recommend you calibrate your sensors at least once every three years.

Powered sensors

Powered sensors are sensors that are actively powered.

Sensors that fall within this category are CO2, airflow, pressure, and humidity. A powered sensor will get its power from a transformer. Depending on the sensor, the power source can be either AC or DC voltage. The power supplied to the sensor is typically in the range of 16-30 volts.

Powered sensors work by using a sensing element that is kept active via the source voltage. The reading from this sensing element is sent back to the controller using a variety of output voltages (0-10VDC, 0-5VDC, 2-10VDC, and 4-20mA). On most sensors, this output setting is controlled by a switch on the back of the sensor.

The output voltage is configured within a field controller using a scale. An example of this is a relative humidity sensor that has a 0-10 Volt DC output.

The 0 VDC equates to 0% relative humidity, and the 10 VDC equates to 100% humidity. This means that for every 0.1 VDC change the relative humidity has changed 1%.

Circuit or switch sensors

The final sensor type that I will cover is the circuit sensor.

What is a circuit sensor you ask?

Have you ever seen a fan status switch?

What about a low-temperature switch, commonly called a freeze-stat?

These are all circuit switches, and they exist to break or complete the electric circuit that is wired back to a field controllers input. In some cases, the BAS technician will run the fan command or damper command from the field control through a switch.

An example of this is a damper end switch. A damper end switch will make sure a damper is open, before letting the fan command reach the fan.

Common protocols that BAS talk

Hello, Hola, Konnichiwa, Hallo.

Across the world, people speak different languages. Fortunately, we have Google Translate to ensure we understand one another. However, this wasn't always the case. We used to have to have translators accompany us on meetings with folks who didn't speak our language.

This person would act as a translator for both groups of people.

For all the great strides that the BAS world has made, BAS communications are still in the midst of coalescing around a standard protocol.

Before I dive into each of these protocols, it might help if I defined what a protocol is. A **protocol** is a standard used to define a method of exchanging data over a computer network.

When your BAS communicate to one another, they are using protocols to tell them how to talk to each other in a way that they will understand.

There is a ridiculously large amount of protocols in the market right now. I am only going to cover the ones that you will regularly encounter as you design, purchase, install, and support building automation systems.

Before I dive into protocols, I want to address a concern that a lot of building automation professionals are facing.

As I said at the beginning of this book, I wrote this book in collaboration with several building automation professionals. When I asked them how to best address this chapter they suggested that I first define what an open protocol is.

After all, no one wants to be locked into a single system right?

What is open?

When I went back and brainstormed what open truly meant I came up with these three questions:

- Is the protocol a National or International Standard?
- Is the protocol interoperable? And, is there an organization that ensures that the protocol fulfills the National Standard and that all manufacturers work with them?
- Is there a minimum of 5 BAS or equipment manufacturers who work with the protocol?

Based on these three criteria I've listed out the following protocols as open. Sure there are other protocols that are open as well but the following protocols have a track record of being open and reliable.

BACnet

BACnet is a wired protocol that communicates across several different network "types." In the world of building automation, the predominant network types are Internet Protocol (IP) or Master-Slave Token-Passing (MS/TP).

BACnet/IP

BACnet/IP used to be solely utilized at the supervisory device level. However, the increasing use of IP field controllers is causing BACnet/IP to be used at the field level.

BACnet/IP is interesting because it communicates using broadcasts to locate devices. However, broadcasts are prohibited from leaving local networks. Because of this BACnet/IP has to use a device called a BACnet Broadcast Management Device (BBMD). A BBMD's sole purpose is to keep a list of other BACnet networks in case a broadcast needs to be sent to a BACnet device on another network.

BACnet MS/TP

BACnet MS/TP is predominately used for field controller communication.

MS/TP utilizes what is called a token ring network for communication. A **ring network** is a network where each controller is connected to its neighboring controller via a 3 or 4 wire cable. This cable forms a line, or daisy chain, as each controller connects to its neighbor.

In an MS/TP network devices communicate by sharing a special message called a token. To talk a controller must request a token. When the token is available, the controller can send its message along with the token to the destination device.

I will go deeper into networking concepts in Chapter 4. For right now all you need to understand is that BACnet has two types of networks.

Here are some quick facts on BACnet:

- **Standards:**
  - International: ISO 16484-5
- **Interoperable:** The BACnet Testing Laboratories (BTL) define the functional profiles and standard variables for the different BACnet profiles.
- **Manufacturers:** Companies like Alerton, Automated Logic, Distech Controls, Echelon, Honeywell, Johnson Controls, Schneider Electric, Siemens, Trend, Tridium, etc.

EnOcean

I struggled as to whether I should include EnOcean or not. Honestly, I haven't seen it used a lot in projects in the US, but I hear it is quite popular in Europe.

EnOcean is a protocol that is focused on wireless communication. What is unique about EnOcean is that its devices use kinetic energy from motion to power themselves. The devices then emit low power radio frequencies to communicate with one another. This communication method is often known as Mesh networking.

Here are some quick facts on EnOcean:

- **Standards:**
  - International: ISO/IEC 14543-3-10

- **Interoperable:** Interoperability is achieved by complying with the EnOcean Profiles and Equipment Profiles which can then be certified by a third party.
- **Manufacturers:** Companies like Delta, Distech Controls, Honeywell, Schneider Electric, Siemens

LONWORKS

Also known as LON (Local Operating Network), LONWORKS is a technological platform based on ISO/IEC 14908 International Standard. LON is normally used at the field level to integrate building systems (lighting, HVAC, security, etc.) to a single open platform.

LON is based on the open protocol LonTalk.

Here are some quick facts on LON:

- **Standards:**
  o   International: ISO/IEC 14908
- **Interoperable:** The LonMark International Association that define the functional profiles, and standard variables that must work each LonWorks product to ensure interoperability.
- **Manufacturers:** Companies like Distech Controls, Echelon, Honeywell, Johnson Controls, Schneider Electric, Siemens, Trend, Tridium

Modbus

Modbus is the language of industrial devices and power meters.

In the building automation space, the term Modbus is almost synonymous with energy metering. Modbus is a protocol that can communicate over IP or RS-232/RS-485(both of which act as serial ports for the protocol).

In the BAS world, the two most common forms of Modbus are Modbus Remote Terminal Unit (RTU) and Modbus Transmission Control Protocol (TCP).

Modbus RTU

In the BAS space, Modbus /RTU is the most common form of Modbus communication. Modbus RTU communicates via a serial port in a daisy-chain architecture. Each Modbus device has a device ID and series of "registers" or points. These registers are often contained in a register file. When a Modbus system is integrated to a BAS, this register file is often added to the supervisory device that connects to the Modbus network.

Modbus TCP

Modbus TCP gives Modbus the capability to transmit data across networks. While most BAS installations utilize Modbus RTU, Modbus TCP can be extremely useful to large-scale BAS implementations. Modbus TCP provides a reliable way to ensure Modbus communication with devices on other networks.

The important thing to know about Modbus is that a device comes with a list of registers. These registers must be mapped into the control system for you to read or write to the device.

Here are some quick facts on Modbus:

- **Standards:**
    o International: ISO/IEC 14908.
- **Interoperable:** The Modbus Conformance Test Program allows manufacturers to self-test or to have their products certified by a third party lab. The test process certifies products for compliance with the Modbus messaging format.
- **Manufacturers:** Companies like Alerton, Honeywell, Johnson Controls, Schneider Electric, Siemens, Tridium

OPC

Open Platform Communications or OPC, is an industrial communication standard.

The primary thing you need to know about OPC is that if you decide to go down the OPC path, you will be paying a premium for devices that are designed for industrial use.

Quite honestly you don't need industrial strength controllers. I've heard several data center and healthcare providers singing praises about the redundancy and processing speed of OPC. These are both valid points, as there are very few BAS field controllers that have out of the box redundancy.

In addition to this, the control loops (programs) that BAS systems run typically make changes to end devices once per second. Industrial devices are made to process data in nanosecond intervals.

In a manufacturing line control loops can require nanosecond response times. Many customers are buying into this "fast" technology only to realize too late that they are often paying a 1.5 to 2x premium for features they don't need.

Here are some quick facts on OPC:

- **Standards:**
  - ○ International: IEC 62541
- **Interoperable:** The OPC Foundation runs a certification and compliance program through its OPC Test Lab. This lab is under the jurisdiction of the Compliance Working Group. The OPC test lab certifies both client and server devices and ensures that these devices can interoperate and support a common set of data types and functions.
- **Manufacturers:** Companies like Iconics, MatrikonOPC

Zigbee

Zigbee is another popular 802.15 Mesh protocol (meaning the network is wireless and self-forming/ self-healing).

Zigbee has become very popular with the new IoT craze going on right now. Zigbee networks use line of site communications to connect devices up to 100 meters apart.

The important thing to know about Zigbee is that it is very sensitive to physical disruption. This means that Zigbee does much better in an open environment like an auditorium versus a hospital with its floor to ceiling walls.

You should realize that the mesh wireless space is still in flux. While Zigbee is popular, it may not be what the industry standardizes on. Because of this, you should be aware of the risk of accepting a Zigbee solution.

Here are some quick facts on Zigbee:

- **Standards:**
  - ○ IEEE Standard 802.15.4 (currently supporting Zigbee 3.0)
- **Interoperable:** The Zigbee Alliance certifies products and platforms for compliance with the Zigbee standard. The alliance also creates profiles for the individual Zigbee products.
- **Manufacturers:** Companies like Carrier, Distech Controls (Acuity), Honeywell, Schneider Electric, Siemens, Trend, Tridium, Etc.

Control logic fundamentals

Controls are logical even if it may not seem that way!

Oh, come on now, you gotta love my play on words :-D.

Seriously, though, at the end of the day, your BAS controls will do what they are programmed to do. Every once in a while you may run into bad physical inputs or outputs, but the control sequences usually work as designed.

So how do controls work?

I mean really work.

Behind the scenes, deep inside your controller, there are several things going on. I could write an entire book on control logic and control modes.

In this section, I am going to focus on the key concepts related to control loops and control modes.

Control loops

This is where the magic happens folks.

All of the parts and pieces, time, spent programming and wiring, none of it matters if the control loop doesn't function!

But what is a control loop?

If you've been in the field for any amount of time you've probably heard the term control loop thrown around, but do you understand what a control loop is?

Well, get ready, you're about to find out!

Here's the best way I can describe control loops. I've gained a couple of pounds since I left the military and that all has to do with me eating more calories than I've burned.

You see there is a limited amount of calories that my body can burn each day. This is my body's setpoint.

But you see I've got a problem.

I love food and the amount of calories I consume, is higher than the setpoint my body has. Because of this, the calorie burning process cannot keep up. This means that my body produces fat and I gain weight.

That my friends is a control loop.

Because my food input is greater than my food burning set point and so my weight increases. Now all joking aside, what does that look like with BAS?

Imagine you have a Hot Water System.

Your BAS is set to supply hot 140° water to the building. As part of this hot water loop, you have a hot water supply and return temperature sensor and a hot water mixing valve.

As your temperature drops below your set point, your "heating process" also known as your control loop increases its output. This causes your mixing valve to open up. Once the temperature goes back above your setpoint, the control loop causes the mixing to close.

There are two types of control loops you need to understand.

These are open loops and closed Loops.

Open loops

I mentioned earlier that there are two types of loops. The first type is the open loop. An **open loop** takes an input and executes a process without providing feedback to that process.

If that was Greek to you, let's break it down a bit.

Imagine you have a warehouse. Inside that warehouse is a bunch of heaters mounted on the ceiling. You know what I'm talking about, those giant ugly gray boxes with a fan and electrical heating coil inside them.

Well, when the temperature in the warehouse drops below a certain point those heaters are enabled which turns on the fans and the heat. There is no feedback to the control process to tell the heater to back off because the space is warming up.

The heater just runs until the temperature goes back above the setpoint.

The key point to remember with open loops is that there is no feedback mechanism!

Closed loops

You must be thinking, well if open loops lack a feedback mechanism then closed loops must have one right?

Damn, you're good!

Seriously, though, that's it! **Closed loops** are loops that have feedback to the control process.

The conversation in a closed loop goes something like this, "Hey Mr. Process, we're getting close to setpoint you need to back-off a bit."

In a closed loop, the control system will calculate the difference between setpoint and input (often called the process variable or PV) and will feed this into the process. This difference is called the **error**. The output of the controls loop will adjust based on the error.

Let's look at an example.

Imagine that you are running a chiller to produce chilled water. Your chilled water supply setpoint is 42° but your chilled water supply temperature is 48°.

That's a problem!

No biggie right?

Your chiller kicks on and starts the refrigeration cycle (you read the HVAC chapter right?). The chiller starts to cool down the water, and the fact that the temperature is lowering is feedback into the control loop. Your chiller notices that the chilled water temperature is being reduced and starts to throttle back its compressor reducing the amount of chilled water that is being produced.

This is a closed loop.

As the input changes, the process adjusts its output. This is because the process is getting direct feedback from the input.

That is why you will often hear closed loops called feedback loops.

But what about these processes?

How are they controlled?

Well, my readers, that is covered in the next section!

Common control modes

Control modes are a key aspect of building automation systems that are often ignored.

I can tell you from experience that one of the most common errors with a BAS has to do with the wrong control mode being used.

For example, do you think that you should use binary, floating, or proportional control with a VAV box?

If you don't know the answer, then you will want to read this section.

By the end of this section, you will know what each control mode can be used for.

Control modes exist to control outputs. **Outputs** in this section are defined as the physical connections that control devices. Some examples of this would be dampers, fans, valves, etc.

Binary control

Binary control, or on/off control is one of the most common control methods in the BAS space and for a good reason. Almost every HVAC system has an on/off device.

In **binary control,** your control system is telling a device to turn itself on or off. Seems simple right?

Even though it is simple, on/off control can cause physical damage to your systems if it's done wrong.

What do I mean by this?

Think about an air handler.

Once again, you did read the HVAC chapter right...

In this scenario, air handler gets air from an outdoor air duct. To get the air, the dampers for the air handler need to be open. Well, this could pose a problem with straight on/off control.

How so you ask?

The problem is that sometimes folks will command components without knowing the status of other components. Back to the AHU example, what if you didn't know the status of the damper before turning on a fan?

You could blow up your duct work. That is why you will often see most binary control modes being physically dependent on a safety circuit.

This means that the fan command would be physically routed through a damper end-switch (this is a switch that closes when the damper is open). Only when this switch is closed would the fan be enabled.

The applications that commonly use binary control are:

- Lights
- Fans
- Pumps
- Unit heaters

Sequenced control

Sequenced control is driven by a software output. The control process will then drive a series of physical outputs.

When the output to the sequencer increases the sequencer will "stage" on new physical outputs. Once an output turns on it is required to run for the minimum on-time. Once the output is turned off it is required to stay off for the minimum off-time.

Sequenced control is typically used to control direct-expansion (DX) units and to stage on and off multi-chiller/boiler central plants.

Floating control

There is a variation of sequenced control called floating control. Floating control is also known as incremental control.

Now some folks could argue that floating control is its own control mode and I would agree. However, to make this easy to understand, I wanted to bundle up related control modes.

Floating control is a common control mode that is used with floating actuators.

I briefly discussed the physical aspects of floating control earlier in the BAS components section. In this section, I am going to discuss the software aspects of floating control.

Floating control operates by commanding two different outputs. You have an increment output and a decrement output. Now, what do I mean by that?

Basically, you have a software output that says your control device (damper, valve, etc.) needs to be at 60% output. Based on this output a sequencer will command the increment or decrement outputs to drive the control device to that %.

You will need to program the stroke time for the actuator, so the controller knows how long to command the outputs to open or close the actuator. The **stroke time** is the amount of time it takes the actuator to fully open.

Pretty simple?

Proportional control

Proportional control is an important control mode.

Why is that you ask?

When you stop and think about the control methods, we've talked about up to this point you've probably noticed that they are dependent on a software input driving their output. For example, sequenced control works by converting a proportional signal into a series of binary outputs.

Proportional control modes consist of two modes, and these are proportional only and proportional-integral (PI) control.

Note I did not include PID control because that is only used in a few sequences.

In **proportional control,** the output is proportional to the difference between the set-point and input. As I mentioned earlier, the difference between the set point and input is called the **error,** and this error is used as the output.

Proportional-integral (PI) control

Most BAS applications use proportional integral control. This mode is used because it introduces a "time" factor to the control loop.

Let's unpack this.

Imagine you have an air handler and this air handler is controlling the duct static pressure using a variable frequency drive. If you only used the proportional difference between the pressure set point and pressure value, you might not adjust the fan speed fast enough, or you might adjust it too fast.

As you can imagine, either scenario could be very bad!

So, what's a BAS person to do?

Well, that is where the integral portion of PI control comes into play. Integral control adds a time variable to your control loop. Each time a PI loop cycles, meaning each time the code runs (usually once a second), the integral portion will look at the error and if the error is still present then the integral value will be added to the output.

The integral value continues to increase, which increases the output.

As you can imagine, you need the right integral values to make sure that the PI loop doesn't "wind up." When a loop keeps increasing or decreasing past 0% or 100%, the loop can become "locked." This is where loop tuning comes into play.

When you tune a loop, you are watching how the PI controllers output responds to the error. When you first install a BAS controller, you will want to adjust the "tuning" factors based on the reaction time of the loop. That's as deep into PI loops as I am going to go in this book.

Chapter 2 Quick Summary

Building automation systems are complex. There's no way to get around that. But, you now have a fundamental knowledge of BAS that quite frankly some BAS techs don't even have.

Let's recap what I just went through.

A BAS supports four main capabilities. Those capabilities are:

- Alarming
- Command and control
- Monitoring
- Trending

Of those four capabilities, the two that get messed up the most are alarms and trends. Alarms and trends need to be thought through before they are setup. Once alarms and trends are setup, they are stored locally on supervisory devices and then batched or streamed to the database. The server will coordinate database storage and retrieval.

Speaking of the server, the BAS server is the brains of the whole operation. This server manages:

- Data storage
- Monitoring
- Multi-site management

However, the server is just one part of the four tiers in a standard BAS architecture. Those four tiers are:

- Server
- Supervisory device
- Field controller
- Sensor layer

Let's be real, a lot of folks have multiple BAS at their sites and at the end of the day, folks just want these systems to work! As I discussed, BAS use open protocols to communicate to one another. It is important to understand the different protocols and the options they give you!

Ok, folks, there you have it. You've got the fundamental knowledge of BAS. But get ready to have your mind blown because there's a whole 'nother world of systems waiting for you in the next chapter.

CHAPTER 3

OTHER SYSTEMS THAT YOU MAY ENCOUNTER ON YOUR JOURNEY

What's in this chapter?

Throughout your career, you are going to encounter a lot of different building systems.

This chapter will expose you to the most common systems you will encounter. As you read through each section you will discover:

- What the system is
- How it functions
- How the system ties into a building automation system

Here are the key learning objectives for this chapter.

- Expose you to the key systems you will encounter
- Help you to tie the functions of these systems back to your business
- Point out common ways these systems can tie into your BAS
- Identify common pitfalls and gotchas

Contrary to popular belief there are systems other than building automation systems, I know, I was shocked as well.

However, the cool thing is that these systems are all starting to come together. If you are just starting to get your head wrapped around the concept of building automation systems, the consolidation of systems in smart buildings can be very frightening.

In this chapter I am going to give you the quick and dirty details on the other kinds of systems you will encounter along your way. Buckle up, keep your hands and feet inside the vehicle at all times because we are about to go for a ride!

Smart building system types

In the following sections, I am going to discuss the systems that you will typically find within a building. While your current building may not have all of these systems, there is a good chance that you will run into all of them during your building automation career.

Audiovisual

Audiovisual (A/V), is a pretty darn self-explanatory term right? Well, this little technology can nip you in the butt real fast if you don't understand what it is.

What is it?

Audio Visual systems are similar to building automation systems in that they have a server, supervisory device, and field controller. An **audio visual system** provides input and output control of audio and visual equipment.

However, A/V systems can be challenging when their integration and use cases are not properly designed.

How does it function?

A/V systems allow the user to monitor and control sound systems, lighting, projection devices, and digital signage.

Often there will be preset "scenes" that will allow the end-user to adjust the space for a predefined scenario. This is the first place where I've seen end-users get tripped up. If there is not alignment between the lighting, audio, and A/V vendor, there can be issues with system performance.

Control of the A/V system is usually provided through a wall-mount controller that will allow the end-user to select a scene or change individual room settings. Newer A/V systems support tablets and mobile devices. This allows users to change the room settings without having to use a wall-mounted display.

The room controllers communicate back to a centralized supervisory device that feeds into a server.

Codecs, API's and other stuff

A/V systems transfer a lot of data. Most of the data they transfer is video and audio data. Because of this, A/V professionals have to deal with both information technology (IT) and information communication technologies (ICT).

Some of the specific things that you should be aware of if you are getting involved in an A/V project is:

- What kind of data will the A/V system be transferring?
- How and where will the A/V system be transferring this data to?

With the advent of 4k video and live-streaming, the amount of bandwidth (network capacity) required for A/V has greatly increased.

In addition to this, you need to be aware of the API's and Codecs that are required. Application Programming Interface (API) allow the A/V system to interact with and control various projectors, screens, lights, and cameras.

Codecs allow the A/V system to process and render different video streams.

How does it tie into a BAS?

A/V systems tie into a BAS at the server or master controller level.

There are two ways that A/V systems "tie" into building automation systems.

The first way is via a BAS protocol, the protocol that is most commonly used is the BACnet/IP protocol.

The second way an A/V system can tie into a BAS system is through an Application Programming Interface (API). An API defines how systems communicate. Think of an API as the grammar rules for networked applications.

When purchasing an A/V system, it is important to evaluate its connectivity options as not all A/V systems are the same. The systems I have had the most success with in the past have been AMX and Crestron, both of which have BACnet/IP interfaces and API's.

While I will discuss integration in great detail in Chapter 15, it is important to note that the data and communication portion of the integration is just one piece.

As you will see in your career, and I will point out repeatedly in this book, the use case should be your primary factor in integration development. When you work with any system, you should, or your designer should identify which system will have which role.

A few things to think about in regards to A/V integration are:

- Should the A/V system or the BAS have primary control of lights, shades, temperature, etc.?
- What priority level is given to which system?
- Who has the ultimate control in regards to overrides?
- How will you handle preferences?
- Will the A/V system track user preferences (temperature, lighting, etc.) or will the BAS?

Conveyance control

What is it?

Conveyance control systems control conveyance systems, man I'm on a roll for making obvious statements. **Conveyance systems**, also known as, elevator systems move things and people around. They do this using belts, cabling, or in some rare instances magnetic-levitation (maglev).

How does it work

Conveyance

As I said earlier, conveyance systems move things around. These systems are typically found in the transportation and industrial spaces. These systems consist of belts that are automated based on preprogrammed sequences. These systems can get quite complex depending on the use case.

Elevators

Elevators are controlled by a centralized controller that is wired to a series of buttons on each floor and in each elevator. The elevator controller then controls a motor to that moves the elevator around. The controller also has inputs from the fire alarm system to override or disable the elevator in the case of a fire.

How does it tie into a BAS system

In the building space, there are two main ways that conveyance systems tie into the BAS.

The first is quite simple, and that is direct wiring. In this integration, status and control points are wired back to the BAS field controller.

The status points are used to tell what floor or location the conveyance system is on.

The next form of integration is protocol integration.

In the case of conveyance systems, the most commonly used protocol is Modbus/RTU. As you may recall from Chapter 2, Modbus/RTU communicates across a serial interface like RS-232 or RS-485. Conveyance devices will typically be wired in series and will be wired back to a Modbus gateway which is connected to the BAS.

Digital signage

What is it?

Digital signage is a system that allows you to display signage, digitally. Done!

Seriously, though, **digital signage systems** use monitors or screens to display information from a centralized server. There are two types of digital signage, passive and active.

A **passive signage** system displays information from a content server.

An **active system** displays content from an interactive web server that allows users to interact with this content.

Digital signage systems are typically used for maps, informational kiosks, and tenant engagement.

At of the time of this books publishing I have noticed a significant uptick in the amount of digital signage being installed on projects. I believe this is largely due to the price of monitors and televisions dropping substantially.

How does it work?

To configure a digital signage system, the customer will log into the server or the individual device using a web browser and IP address. They will then be able to configure what is displayed. Typically this is a static image, PowerPoint, or website.

In some cases, applications are developed and pushed to the screens. If these are touch screens the users can use the screens to interact with the applications. Examples of this would be interactive kiosks, maps, and directories.

Some of the places signage is used at are airports, hospitals, and stadiums.

How does it tie into a BAS?

This is where the area of A/V and signage blur slightly. Technically, several A/V vendors have signage solutions, but these are often provided as part of a greater A/V package. If you are ok with a more integrated signage solution, then a solution from an A/V provider will provide you the same integration capabilities via BACnet/IP or API's.

This will, in turn, allow you to display your BAS data on a signage solution.

Examples of this would be the Crestron 3-series.

Now if you want a dedicated signage solution you have a slew of potential technology providers. While this provides you several advantages in the area of competitive pricing and availability, it also creates issues around integration. The primary issue stems from the lack of integration standards in the signage space.

The most common integration method you will see from a signage solution is the ability to use web programming languages, like JavaScript and Angular/JS to call against BAS API's to display HVAC data.

As you might have guessed this means you need to account for the costs of a software developer when tying into these systems. This risk is quite common with several of the non-BAS systems in this chapter and one I want you to get good at looking out for!

If you take one thing from this chapter, it should be to ask yourself constantly,

"How will these systems talk to one another?"

Now understand, you do not need to know the answer! You simply need to ask the question and let others answer it. So many of the issues I've experienced on projects stem from not asking how the systems will talk to one another.

Fire alarm systems

What is it?

Fire alarm and fire suppression systems are a key component of nearly every system. Whole books have been written on fire systems, and I by no means claim to be a fire system expert. I am going to tell you what I believe you need to know based on my experience in regards to building automation systems and fire systems.

First, off the interface between fire systems and building automation systems can be split out into two buckets smoke/pressurization control and fire alarm safeties.

How does it work?

Smoke control/pressurization

Smoke and pressurization control are where the building automation system has been programmed to execute a specific sequence in response to a fire or other environmental hazard. In the case of a life safety event, the building automation system will be notified by the fire alarm system. This is typically done using a hardwired signal. This signal will be wired to a BAS controller that will let the BAS know to command the HVAC systems to perform a pressurization or smoke control sequence.

Fire alarming

Fire alarms systems are centralized systems that receive signals from remote sensors and switches that allow the system to sense a fire. Once the fire alarm system has sensed a fire, it will send a signal to the building automation system which will respond based on its programming.

In addition to this remote signal, some HVAC equipment will have fire and smoke detectors that are connected to the nearest building automation controller.

How does it tie into a BAS?

For fire and smoke control systems the most common form of integration I have seen used is direct wire connections. In most cases, this is done to satisfy code requirements for the local municipality. There are typically two different points that are wired back to the BAS system.

The first point is the general fire alarm. This point is typically wired back to a controller within the vicinity of the fire alarm system.

By the way, I recommend tying the fire alarm status point into a separate controller. If you don't do this and you have to replace the field controller or reprogram it, you risk losing your fire alarm status.

In some cases, the general fire alarm status is wired to the controllers for all of the major HVAC systems.

The second point that is commonly wired up is at the actual field controller. This point is a local smoke or fire alarm switch. This switch is either wired directly to the controller as a dry contact, or it is used to break a control signal between the control and a device, typically a fan.

Lighting systems

What is it?

Lighting is so complex that I could write a whole book on it.

In this book, I am just going to give you the basics of what you need to know. Lighting consists of a power source, a lighting fixture, and a switch to allow that power sources to reach the lighting fixture.

Now a brief segue on lighting litter for my friend Kenny Smyers

There is a trend right now around a concept called lighting litter. Lighting litter is a term used to describe when light is being wasted. A good example of this is using exterior lighting during the day.

This is a huge waste of energy that can be easily resolved through a local controller that schedules the lighting systems.

Back to lighting systems, lighting systems as I mentioned include a power source, fixture, and switch.

How does it work?

If you were to look at a traditional lighting system, you would see three parts that make up the lighting system. Those parts are the power source, the fixtures, and light switches. Sure there are other parts like occupancy sensors, daylight light sensors and external control devices like shade motors but those are auxiliary pieces.

Power source

In traditional lighting, the lighting system is powered by line voltage. **Line voltage** is voltage that is 120 volts or higher. Individual lights are usually controlled via an on/off switch. A slight variation of this is to use a slider, also known as a **dimmer**, to control the light level.

Notice, I haven't started to discuss actual lighting control systems yet.

Fixture

The **fixture** contains the actual light and typically is installed in the ceiling (indoors) or on a poll or mount (outdoors). There are a variety of fixture types. Going into the individual fixture types is outside of the scope of this book.

Light switch

The **light switch** allows the user to turn the light on and off or to adjust the light levels. In some cases, the light switch will send a control signal (0-10 volts direct current) to the light. In other cases, the light switch will "break" the connection between the power source and the light fixture.

Some newer light switches are wireless.

Lighting control systems

Next, up, I have lighting control systems.

Lighting control systems take the concept of lighting a step further, by introducing an automated control system that oversees the control of the lights. Just like A/V systems, a lighting system architecture looks very similar to a building automation system. This is because a lighting system has a supervisory controller, individual zone controllers, and in some of the newer technology, there are controllers at the individual lights.

I want to take a second to point out something that I hope you are noticing. The majority of these systems (A/V, Elevator Control, Lighting, etc.) all function the same way. This is because most systems are built on the supervisory device => field controller => input/output architecture. Once this clicks for you, it will make it much easier to understand how building systems work.

Ok, back to lighting, lighting systems are often set up to control a series of lights in a "zone." A **zone** is an area that can stretch across multiple rooms and is usually designed to serve a specific purpose. For example, you may have a zone that is a general lighting zone, and then you may have a zone that is designed to cover all the lights in the individual offices.

This brings up one of the trickier areas of lighting integration, and that is how do you reconcile "zones" to individual spaces?

Here is what I mean by that. In the BAS world, we often have a one-to-one relationship between our field controllers and the space they control. Now, granted you may find some buildings that have the diffusers for multiple spaces connected to a single VAV box.

Ok, so what does this mean to you?

Well, if you have a lighting system that is controlling lights for a zone and this zone stretches across multiple spaces, you can run into a scenario where you have a many-to-one relationship.

When you go to integrate your lighting system to a BAS, you can't automatically pair the lighting zone to the individual spaces. This ends up causing you to spend a considerable amount of time going and creating a "middle layer" that maps the individual BAS spaces to the lighting zones. If you don't plan for this, you can end up spending a significant amount of time doing this point mapping!

PoE Lighting

Power over Ethernet (PoE) is invading the BAS space. I'm still not sold on its value, but I have seen it pop up on multiple projects. One of the ways this technology is being used is to power LED lights. In my professional career, I am working on several PoE lighting projects.

PoE is not a new technology.

What is new is the use of PoE in BAS and lighting systems. This trend within the building automation market allows devices to be powered over Ethernet cables using low voltage DC power to power devices.

This approach allows the end device to be powered by low-voltage wiring that draws significantly less power and is much easier to setup. To illustrate how much easier, it is to setup. I had the lab gear for a specific PoE lighting product, in my basement and my 8-year-old son and 10-year-old daughter were able to set up the PoE lighting in an hour by following the installation document.

This has major ramifications on the cost of installation and the cost of lifecycle support. Although PoE lighting products may not be at a cost point that is efficient for the average project to adopt right now, they will be in the very near future.

How does it tie into a BAS?

Oh man, this is a loaded question...

The lighting market is going through the same shift that the BAS world went through back in the early 2000's. There are four main protocols on the market that are vying for dominance, these are:

- CoAP
- Dali
- Mesh (ZigBee)
- Wi-Fi

With that being said I'm going to focus on the main integrations that you will see. At the time that this book is being published those technologies are BACnet/IP, Dali, ZigBee and direct wiring.

So how do you integrate each one?

In the case of lighting systems, the first thing you should consider is what capabilities do you need?

For example, if all you want is the ability to turn on and off all the lights in a building, then you could probably get by with just using direct wiring. On the other hand, if you wanted to go and have control and monitoring at the individual fixture level then you would need to consider an integration.

Now comes the fun part! Before you purchase a lighting system, you need to sit down and evaluate the lighting manufacturers that you prefer and then you need to find out the following things:

- What protocols do they support?
- What points and functionality do they expose across their integrations?
- Do they support prioritization?

What protocols do they support?

What you are looking to find out here is what are the protocols that the lighting manufacturer supports that your BAS also supports. You also want to find out how they implement these protocols. For example, do they implement the BACnet protocol via a server or an integration card? This makes a big difference when you are trying to plan out your network architecture.

What points and functionality do they expose across their integrations?

Next, you need to find out what data and functionality the lighting system will provide you through their integrations. Most importantly you want to make sure that the use case you are trying achieve can be fulfilled with the data points and functions you are receiving from the integration. In my experience, this is the part folks screw up. If you get this part wrong, you can pretty much bet on spending anywhere from 3-5x your install cost to fix your mistake.

Now, I can see some heads turning as that is a pretty bold statement, but I've seen entire lighting systems replaced because they didn't provide the appropriate functionality.

If you follow these steps, you can greatly reduce the chance of this occurring. These steps are:

- Detail out the points you need exposed from the lighting system and include those in your RFP
- Detail out your sequence of operations and clarify which system will be executing the sequence.
- Require the manufacturers to respond with how their system will execute these sequences
- Ensure that the lighting system supports prioritization if it will be controlling the same points as the BAS

Do they support prioritization?

Prioritization is one of the commonly misunderstood areas of BAS control. Don't worry, though, when you finish this section, you will be well-versed in the art of prioritization foo!

When two systems need to talk to the same point a delicate dance occurs. What you don't want happening is for the two systems to override one another repeatedly. To ensure that the system you decided would be the main system takes the lead you need to implement a prioritization strategy.

So, guess what I am going to cover next?

You guessed it a prioritization strategy!

Recently, I was working with a lighting system to link the lights to individual user profiles. In this use case, a user profile would include the temperature preference, light level preference, and sound masking preference. When the user badged into the building, the systems would adjust based on these preferences. The problem was that the lighting system was setup with five different themes or lighting profiles.

Now how, could I tell the lighting system not to change this individual space according to the lighting themes when my use case was running?

I did this through a concept called a priority array. Now some integrations don't support prioritization, but mine did. I first looked at the points I needed to control, and then I made the matrix you can see below.

	BAS System Use Case	Lighting System Use Case
Light On	Priority Level 14	Priority Level 15
Light Level Command	Priority Level 14	Priority Level 15

Now at first glance, this looks like a relatively simple matrix.

The thing is as I stretched this across multiple zones and systems the matrix got rather complex. However, with this matrix, I was able to see what system has priority in each use case.

Side note, for those of you with a BACnet background you will notice that I used a lower priority and I did not use priority 8 (this allows me to perform operator overrides on the BAS use case). In the BACnet world, the lower the number, the higher the priority.

Now that I have the prioritization matrix laid out I can do one of two things.

I can include this matrix on my RFP or specification and submit it as part of the project requirements. Or, if I have a direct vendor relationship with my lighting and BAS partners, I can bring them to my site and have a meeting around how we can execute this prioritization strategy.

That's it, most of my customers tell me that displaying my matrix this way makes it very simple to understand. My hope is that by discussing this, you feel much more comfortable with how to address prioritization issues on your projects.

Maintenance management systems

What is it?

Maintenance management systems (MMS), not surprisingly are used to manage maintenance.

These systems are often blended with work order management systems to process preventative and proactive maintenance. There are a variety of MMS systems on the market right now, and it can be confusing as to what system you should choose.

There are three key features of a MMS solution that you must consider:

1. Work order tracking
2. Equipment and Asset database
3. Reporting capabilities

Work order tracking allows the user and administrator of the MMS to track the work order from the moment it is issued to the moment it is completed.

The **equipment and asset database** allows the user and administrator to track building assets from the day they are procured to the day they are replaced.

The end user can also combine the **reporting capabilities** of the MMS with the work order tracking and asset databases to provide a single historical view of asset performance.

These capabilities allow the site owner to make choices on equipment and maintenance procedures based on past performance.

How does it work?

Typically, a MMS consists of:

- A web-browser based interface that is used to access the system
- A central server to run the MMS application
- A database to store information on maintenance activities and assets

Once the user is logged into the system, the user can see any service requests or work orders assigned to them. Some MMS also provide the capability to route work orders and service requests to individual users.

I want to take a second and explain the difference between a service request and work order. At first glance, it would appear that a service request and work order are the same thing, but there is a key difference between the two.

That difference is who pays.

A **service request** is a request outside of the normal maintenance or equipment repair activities. An example of a service request would be moving a thermostat or the unplanned installation of a new HVAC device. In this scenario, the cost would be borne by the business unit that is responsible for the building.

Work orders are regularly planned maintenance activities or asset repairs that are required to maintain the building in it's as-built condition. In this scenario, the costs would be borne by the facility team.

Back to MMS.

With an MMS, users can also view each piece of equipment and any maintenance that has been performed on or assigned to that piece of equipment. This provides a historical snapshot of the asset.

Finally, users can run reports on a variety of metrics. Some of the more common metrics are:

- Current outstanding work orders
- Work order completion rate
- Equipment maintenance compared to a mean average

How does it tie into a BAS?

This is one of the areas where MMS seem to struggle. It has been my experience that there is no accepted standard for integrating with a MMS. The three main forms of integration I have found are:

- Application programming interfaces (API)
- Protocol integrations BACnet, Modbus, LON
- Self-developed data adapters

Application programming interfaces (API)

Application programming interfaces or API's are used to connect two applications. In the case of a MMS, the API often comes from the MMS and is consumed by the BAS. In this scenario, the BAS will read the API from the MMS and will send messages to the MMS in the format it requests. The most common use of a MMS to BAS integration is to route alarms into the MMS for work order management.

I have found that the main challenge with this type of integration is managing who bares the cost of these repairs. In my opinion, this is one of the key reasons that there has been limited adoption of automating alarms into MMS.

Protocol integrations BACnet, Modbus, LON

Protocol level integrations via BAS protocols into the MMS are becoming more common within the newer MMS. In a protocol integration, the BAS will define the protocol standards and the MMS will adhere to the messaging format that the BAS supports. This interface still serves the purpose of routing BAS alarms to the MMS.

Self-developed data adapters

Self-developed data adapters are coded by the owner, controls company or third-party developer. A data adapter takes the messaging format of the MMS and adapts it to the BAS system. These are the hardest integrations to deal with as they are often not supported by the MMS or BAS provider.

Metering

What is it?

Metering, logging, recording.

Multiple words that describe the same thing. At the end of the day, it's all about logging values. Meters come in multiple shapes and sizes, and if there's anything I've learned in my years of working with them it's that they can be damned confusing!

In this section, I am going to add some clarity to meters.

How does it work?

There are a variety of meter types. There are so many that I am just scratching the surface of the metering "ecosystem." In my experience, the meters covered below represent 80% of the meters that you will encounter in your career. The two most common ways in which meters communicate are:

- Electrical signaling
- Mechanical feedback

The first way is electrical signaling. An **electrically signaled meter** will communicate the sensed value back to a control device using an electrical signal. Often called, **proportional signaling**, these signals will scale the electrical signal proportionate to the sensed value.

An example of this would be a meter that senses relative humidity. If this meter used a 0 to 10VDC output, it would adjust 0.1 VDC for every 1% change in relative humidity.

The second way in which a meter can communicate is called **mechanical feedback**. In this method, the meter will cause a mechanical change to be executed based on the sensed value. Examples of this would be temperature gauges and pneumatic controllers. In each of these cases, the mechanical force of the variable that is being measured will cause the gauge to change the value.

Meters that use mechanical feedback are often used for local reporting and are rarely tied back into a BAS.

For the most part, meters work by providing an electrical signal to your field controller or supervisory device.

Electrical

Electrical meters are the most common meter type you will see inside a building. They usually come in 3 types, in-line meters, multi-circuit meters, and smart meters.

In-line meters

In-line meters are meters that are in-line with devices. I don't how else to describe that without using the term in-line. This meter type is typically a one-to-one meter. This means that the meter senses voltage or power for one data point. An example of this would a meter designed to detect the voltage coming into the building or coming into the system.

Multi-circuit meters

While in-line meters are used to monitor the power of a single device a **branch or multi-circuit meter** is used to measure multiple devices or circuits. A common example of this would be a Veris 24 circuit 3-phase meter that allows you to meter 24 single phase (meaning single wire) circuits and one incoming three wire circuit.

Smart meters

Smart meters are used to measure the power on single or multiple circuits. The big difference is that the smart meter can communicate with other meters or to communicate with centralized meter monitoring software. These kind of meters are very common for utility meters and are becoming more common for commercial meters.

Fuel meters

Fuel meters are one of meter types dedicated to measuring liquids. I've usually found fuel meters on the supply lines going to and from the fuel tank. These tanks are typically used to store fuel for a building's generator. Generators are used in the event of a power outage.

Gas meters

Similar to fuel meters, **gas meters** exist to measure the gas going into a building or system. The utility usually provides gas meters that reside outside a building. The owner usually provides gas meters inside the building.

Liquid meters

Liquid meters is a catch-all term that I reserve for other uncommon meter types. Examples of this are condensate meters, liquid hydrogen meters, etc.

Sewage meters

Sewage meters are utilized to measure the sewage that is exiting a building. You will often see these installed by the water company to collect sewage production for appropriate billing.

Steam meters

Steam meters are used to measure steam.

What a surprise eh?

Steam meters are very similar to gas meters. Steam meters can come in various types, and you need to be careful that you buy the right meter for the pressure level you are measuring. Be sure to pair a steam meter with a condensate meter so that you are measuring the amount of steam being converted to condensate.

Water

Water meters are used in several applications. Water meters are used to measure domestic water, chilled water, irrigation systems, etc. Water meters measure both the amount of water used and the amount of water flowing.

How does it tie into a BAS?

When it comes to tying meters back into the BAS there are three methods that I use:

- Physical connections
- Protocol connections
- Wireless connections

Physical connections

The first and most common method for tying meters into a BAS is to use a direct connection from the meter to the BAS. In this method, the meter has a set of output terminals that an electrician or technician can use to connect a wire from the meter to the BAS controllers input terminals.

Protocol connections

The second way to connect meters back to your BAS is to utilize a protocol connection. Some of the more common protocol connections are Modbus and BACnet. In this scenario, the meter is hooked up to either a local field or supervisory network based on the protocol used.

The meter is then mapped into the BAS. You need to make sure that your environment can meet the meters IT requirements. An example of this would be trying to use BACnet/IP meters when you do not have control of your IT environment and cannot issue IP addresses to the meters.

Wireless connections

The third and final way to connect meters back into your BAS is via wireless sensor networks. The introduction of wireless metering is a newer trend in the BAS world, and while it has its pros, it also has its cons. One of the pros of wireless metering is that you can utilize your existing wireless infrastructure to enable the sensors communications.

However, the con of wireless communication is the fact that the meter communicates wirelessly which is an area that some folks in the BAS world don't yet understand.

In addition to this, there is a perception that wireless is less secure than wired networks. Let me assure you. A properly designed wireless network is every bit as secure as a wired network.

Scheduling systems

What is it?

This may shock some folks, but the building automation system is not the only scheduling system in a building. There are so many different scheduling systems that to list them all would take multiple pages. Therefore, I'm going to break out scheduling systems into two categories:

- Local scheduling systems
- Networked scheduling systems

How does it work?

Local scheduling system

An example of a **local scheduling system** would be a wall mounted time clock. A **time clock** is a clock that activates a relay based on a time setting. The contacts on this relay are then wired to systems throughout the building. The connections from the time clock serve as the occupancy signal for the building. Time clocks are present in a lot of older buildings and are notorious for failing.

Networked scheduling system

A **networked scheduling system** is a system that allows you to set your schedule and then communicate the schedule via an API to other devices (we will cover API's later in the book). Examples of this kind of system are Google Calendar and Outlook Calendar. I've worked on buildings where we took data from the Google Calendar software to schedule the occupancy of conference rooms. It was pretty cool to do.

How does it tie into a BAS?

Scheduling is important.

In an ideal world, the BAS will enable building systems to run only when needed. To meet this demand, the BAS needs to be tied to an accurate schedule. In an ideal world the BAS would be integrated with your scheduling system, but as we all know, things are usually less than ideal.

First off, I am going to look at how the local scheduling system can tie into a BAS. A local scheduling system often does not have a way to communicate to other systems across a network.

Because of this, the BAS must rely on direct connections from the local scheduling system. The local scheduling system usually has a set of outputs that can be connected to a physical input on the BAS controller. Once the BAS detects that the output has been triggered it can then switch the BAS to occupied mode. This form of operation is very error prone and is a non-desirable method for schedule integration.

In newer buildings, the BAS can connect to scheduling systems using the network. The nice thing about using the network to schedule systems is that you can schedule your building in advance. The downside of these systems is that if someone enters the wrong schedule by mistake, you can shut down your building during occupied hours.

I like to set a static schedule on the BAS and then use the network scheduling system to address after-hours usage.

Security

What is it?

In this section, I am discussing physical security. Physical security is a broad topic that includes things like turnstiles, biometrics, physical security devices, etc. Physical security also includes video surveillance and video analytics.

How does it work?

Video surveillance

With all of the craziness going on in the world lately, it's becoming increasingly important to make sure that building occupants are safe. One of the ways that building owners accomplish this is through video surveillance. Video surveillance systems consist of three main pieces:

- Video management server
- Video cameras
- Video storage

Video management server

The **video management server** collects video streams from multiple cameras and allows the operator to view each video stream. In addition to this, the video management server processes video and routes it to video storage.

Video cameras

Video cameras record video. Video cameras breakout into two main categories:

- IP cameras
- Coaxial cameras

IP cameras

IP cameras use Ethernet wiring to stream high definition video. IP cameras can utilize their Ethernet wiring for power as well. As I mentioned earlier in this chapter, this is a concept called Power over Ethernet (PoE).

Coaxial cameras

Coaxial cameras often called coax cameras use coaxial cable to transmit data from the camera to the video receiver. Unlike IP cameras Coax cameras need to be powered using an external source. Typically, they are powered via a 12+ VDC power connection.

Video Storage

Video storage consists of three different types of "storage." These storage types are:

- Solid state disks
- Hard disk drives
- Tapes

Solid state drives

Solid state drives (SSD), allow the video to be recalled quickly and are often used for live video replay.

Hard disk drives

Hard disk drives consist of small "platters' which are thin metal disks that store data for long-term recall. These drives are used for long-term storage.

Tapes

Tapes are used for archiving video. Tapes used to be used for video backup but are now being replaced by hard disk drives due to the price of disk drives being so low.

Card access

Card access systems provide access to spaces. These systems use a card reader and magnetic or RFID cards. The cards will emit a signal that contains an encrypted key. A card reader will receive this signal and send it to a controller where the signal is validated against a database. If the correct signal is provided the controller will command the card reader to unlock the door. Often times card access systems are part of a greater security system.

How does it tie into a BAS?

There are a variety of ways that a security system can be tied into a BAS system. The main ways I have used are API's and protocol integrations.

In the API method, a programmer will write a piece of software to connect the security system and the BAS. This software will receive specific data points, usually binary (on/off) values, from the security system. These values are then normalized and sent to the BAS. I describe this kind of process in greater detail in Chapter 15.

The second method for integration is to utilize the protocol integration method which I have described multiple times throughout this chapter.

Smart equipment

What is it?

The concept of smart equipment could be better described as containerized or packaged controls. Smart equipment is where the manufacturers HVAC units (rooftops units, chillers, etc.) have controllers built into the device and are designed to support remote connectivity.

How does it work?

These devices support a slew of capabilities but the main three are:

- Self-commissioning
- Analytics and Diagnostics
- Automatic sequence adjustments

Self-commissioning

The promise of smart equipment is that you can install it in your building and it will come preprogrammed. You simply need to connect the device to your ductwork, piping, and power. The device then will go through a process, known as **self-commissioning** where the equipment will automatically configure itself and will connect itself to its user interface.

The implications of this technology are **HUGE!** If manufacturers can get this technology to a price point close to normal equipment, then this technology could greatly reduce the need for individual building automation controllers.

Analytics and Diagnostics

Maintenance is one of the areas that smart equipment is focused on changing. With traditional equipment, you wait for a filter switch or pressure indicator to tell you something is wrong with your equipment.

Smart equipment uses a set of predefined "rules" to analyze your systems performance and proactively tell you what is wrong. Smart equipment will attempt to fix any faults it detects.

In addition to this smart equipment seeks to gather information and summarize that information into an intuitive summary page available via a modern web browser.

Automatic sequence adjustments

One of the cool features of smart equipment is the way smart equipment handles the addition of new components. The promise of smart equipment is that if want to add an economizer or a new coil you simply install the device, plug in the extension cable, and select the new feature in the graphical user interface. If this works as designed, it should be massively easier than a traditional retrofit.

How does it tie into a BAS?

Smart equipment is constantly evolving. That is why this next section is quite difficult to write. Depending on the smart equipment vendor you select you could find yourself stuck with a proprietary system, a protocol integration, or an API interface.

The trick with smart equipment is understanding if the system you are integrating with can expose the correct points to your building automation system. This will look different depending on the smart equipment manufacturer system. For example, some smart equipment systems require you to purchase additional communication cards to expose their points outside of the vendor's system.

However, if you follow the standardization process I lay out in Chapter 10 you will have a clear BAS standard to communicate to the vendors that will tell them what their responsibilities are in regards to connecting with your system.

Window Systems

What is it?

When I mention window systems, what is the first thing you think about?

Do you think about automated shades, self-fogging glass, or even touch screens that appear on internal glass walls?

You should.

It's often a surprise to folks when I tell them about the varieties of technologies that are available for window systems.

How does it work?

In this section, I am going to focus on the three systems that are present in most modern buildings. These systems are:

- Operable Windows
- Shade Control
- Reactive Glass

Operable windows

Operable windows are just that, windows that can be open and shut. In several new buildings, windows are being motorized to provide the building operator with the capability of automatically opening and shutting windows based on the environmental conditions outside.

Shade control

Shade control is commonly used with a daylighting strategy where certain parts of the building are shaded based on the thermal load (sunlight) that is being cast upon the building. Shade control is often operated in sequence with the lighting system.

Reactive glass

Reactive glass is a newer concept that is gaining adoption. **Reactive glass** uses an electrical current to cause the glass to either change color or to "fog up." Reactive glass is being used for both privacy and as a way to shed thermal load. I like to think of this as a transition lens for my building.

How does it tie into a BAS?

In the case of window systems, I use the same three methods I've been describing throughout this chapter. Those methods are:

- Direct connection
- Protocol integrations
- API's

Side note, at this point you may be tired of me saying direct connection, protocol integration, and API.

I want you to know that this was by design.

Each time I mentioned an API or protocol integration I was using the concept of repetition to solidify this concept in your head. The idea is that by the end of the book you will be so familiar with these three integration methods, that whenever you have to integrate a system, your mind will automatically consider how you can make the systems talk.

Trust me. This will save you so much money, time and pain over the course of your career

The three methods of integration supported by window systems are direct connections, protocol integration and API's.

With a direct connection, each window controller has an output board that is tied back to a local field controller via a hardwired connection. When the sequence of operations calls for the window system to be operated the field controller will command the window system via its output.

With both protocol and API integrations, the window systems are connected to the BAS system via a network-level integration. This integration allows the BAS to command the window system based on the sequence of operations.

Chapter 3 Quick Summary

Man oh man, is your brain tired yet?

Well good news, we are halfway through the first section, Yay!

In this chapter, I took you on a journey through a lot of the different systems that you will encounter in a building. I love seeing the wonder in folk's eyes when I show them the multiple systems **outside** of Building Automation Systems that they need to be aware of.

I can tell you from experience, understanding how non-BAS systems function will enable you to design, install and manage a BAS. When you consider how many new buildings are being designed with integrated systems, this is stuff you gotta know!

In this chapter, I also gave you your first taste of systems integration. I showed you the three main ways to integrate non-BAS systems with BAS systems. Ok, pop quiz, don't read on until you answer this question...

What are the three ways to integrate systems?

If you answered, direct connections, protocol integrations and application programming interfaces you are correct. Any other answer is wrong, and you need to go re-read the chapter. Don't worry I'll wait right here for you to finish.

Finally, I covered how important it is for you to consider the capabilities of your non-BAS systems when planning your projects. This knowledge will become very important to you as you implement complex projects.

Booyah!

Drop the mic!

But wait, let's pick that mic back up because in this next chapter I'm going to speak to you about something that scares the pants off of folks and that's the topic of IT...

CHAPTER 4

YOU DON'T HAVE TO BE AT WAR WITH YOUR IT DEPARTMENT

What's in this chapter?

There is a greater need for IT skills in the BAS world than ever before. Yet, at this time of increased need the cries for IT training lie largely unanswered.

In this chapter, I will change that. By the time you finish reading this chapter, you will understand what an IT department is and the fundamentals of information technology.

Is that a lot to promise?

Yes, but I've got good news, I've secretly been filling your head with IT concepts in the previous three chapters. Now, it's time to fill in the gaps.

In this chapter I will explore:

- The problem that exists right now between BAS and IT groups
- How an IT organization works so that you know who to engage?
- The fundamentals of IT systems so that you know which ones to use
- A deep-dive into the top 4 methods for remotely accessing your building automation system
- A primer on cyber-security

So, are you ready to take your IT skills to the next level?

Let's do this!

The problem

Contrary to what you may have heard your IT group does want to help you! I talk with IT folks all the time, and they are just as confused by the BAS world as we are of their world.

Despite this, there still seems to be a vicious rumor running around that IT groups exist just to hold back controls companies and owners from accomplishing their goals. I want to tell you that in my experience this is almost never the case!

Sure, in some cases there are IT groups that don't want to do any extra work. However, as I stated earlier, it is my experience that those cases are the exception to the rule.

In my experience, the real reason folks don't want to work with their IT groups is because they don't understand IT.

And you know what, I'll let you in on a dirty little secret…

I've talked to several IT folks, and they feel the same damn way about BAS!

So here's the deal. If you give me 45 minutes of focused attention, I promise you that you will leave this chapter understanding the fundamentals of IT

Does that sound good to you?

If so, I'm going to start by giving you an understanding of how an IT organization functions.

How an IT organization functions

An IT organization consists of a couple of functional groups. These groups will be covered in the following sections.

Now you may be asking yourself, "Why do I need to know this?"

That's a fair question.

The reason you need to know how an IT group functions is so that you know who to go to when you encounter certain BAS issues. That's why after each group's description I've listed out the BAS tasks that each group can help you with.

There is one thing I want to make sure I mention to you. The names I use to describe the groups may be different than the name of the groups you encounter at you or your customer's site. The functions, however, will remain the same.

Functional IT groups

Helpdesk

The IT helpdesk is the role where most IT professionals start their career. The **IT helpdesk** handles the wonderfully mind numbing calls like my computer won't turn on, or I forgot my password. An IT Helpdesk group is usually structured in a 2 or 3 tier system that escalates calls based on the calls severity.

Commonly works on:

- Entry level IT issue resolution
- General IT configuration issues

Network administration

The **network administration** group oversees the network and the devices that allow the network to communicate. When you need a new IP address, or you are having trouble diagnosing issues related to network communications, the network administration group will often be involved.

Commonly works on:

- Providing VLAN's, IP addresses, and routing for BAS systems
- Involved in setting up VPN's and other remote connectivity options for BAS
- Involved in assisting with wireless system setup and design

System administration

While the network administration group is focused on the health of the network and networking devices, the **systems administration** group is focused on maintaining systems.

But, what is a system?

There are multiple interpretations of what a system is. The systems administration group focuses on business and IT systems. Examples of these would be e-mail servers, SharePoint sites, etc.

You will work with the systems administration group to set up virtual servers and physical servers. You will also work with the systems administration group to get plug-ins and dependencies installed on your laptops and other devices so that you can run your BAS software.

Commonly Works On:

- Providing servers for your BAS and virtualizing current BAS servers
- Setting up desktops and workstations for BAS access
- Planning out upgrades and patches for BAS systems
- Maintaining username and passwords if your BAS uses company credentials for authentication

Database administration

The **database administration** group oversees databases. They will be the ones to set up, maintain, optimize, and backup databases. Often you will be involved with this group at the beginning of a project to setup the database for your building automation system.

This group can also be of immense value to you if you need reports run on the data from your BAS. If you find the right person, you can have them setup queries (reports) that can be run automatically against the database. This will allow you to report on the data inside your database.

Commonly Works On:

- Providing a database for your BAS
- Maintaining and backing up databases for your BAS
- Creating "views" to look at the data within your BAS database

Architecture

The **architecture group** is often found in medium to large organizations. This group exists to help create technology standards and to drive consistency across your IT architecture. Think of this group as the equivalent of a building automation standards body. This group exists to help the organization standardize their architecture across four functional areas:

- Business
- Data

- Applications
- Technology

You will often hear these areas referred to as BDAT, which is the first letter from each of the areas listed above.

The architecture group should be involved in your BAS planning efforts, but sadly I've seen very few organizations take advantage of this group. I highly encourage you to involve the architecture group in capital planning and creating a building automation roadmap.

Commonly Works On:

- Helping you map out a roadmap for implementing smart building systems
- Coordinating the lifecycle of your BAS
- Helping you to create use cases for your BAS based on business needs

Information assurance group

Information assurance, often known as cyber-security, has become a hot topic lately. **Information assurance groups** (IAG's) have been formed to oversee the assurance that information is confidential, maintains integrity, and is available.

The confidentiality, integrity, and availability of information for an organization are known as the **CIA triad**. Now don't confuse CIA with the Central Intelligence Agency. The purpose of the CIA triad is to enable the business to perform their daily functions securely.

In my experience, there is an understandable reluctance to engage the IAG both from internal groups and from external vendors. I think this has to do with the fact that there is a language and experience gap.

Well, maybe experience is not the right word.

I'd say it's more of an importance or priority alignment issue.

You see, the IAG group is focused on providing confidentiality, availability, and integrity of data across IT systems. However, the constraints required to achieve this often challenge the functionality of the business.

After all, I'm sure you've heard time and time again from security experts how much of a "No, no" it is to enable your BAS to be exposed to the outside world via an external IP address.

But without this external access, many customers are having to spend multiple man-hours if not man-days physically supporting systems that they could maintain remotely.

This has a large cost to the business, and it's why many BAS's tend to find themselves "magically" having an external IP address. It's also why so many IT folks are reluctant to put a BAS system on their network.

Commonly works on:

- Helping you establish security standards for your BAS
- Evaluating BAS products for security compliance
- Implementing a secure BAS installation and holding vendors accountable

Parts of the IT puzzle

Awesome sauce!

So you now have a basic understanding of what an IT group is, what each functional group does and how they are involved with your BAS.

I am now going to build on that understanding by teaching you the fundamentals of IT.

In this section, I am going to cover the key concepts of:

- IP addressing
- Networks
- Servers
- Databases

You ready? This is going to be fun!

IP addressing what you need to know

Some would argue that information technology all begins with the IP address. I don't know if I'd go that far, but it is an important concept.

There are several reasons why you should understand IP addressing. In my opinion, the most important reason to understand IP addressing is that building automation systems are becoming fully IP based.

What does that mean you ask?

Well quite simply it means that the main way BAS devices will communicate in the near future is via the Internet Protocol (IP).

Which brings us to...

What is an IP Address?

An IP Address is a complex topic. There is so much that can be said about IP addresses, but in reality, there are only a couple things you need to know.

First off, an IP address is an address for a networked device. I am very intentional in using the term networked device. I want you to associate IP addresses with more than just your computer.

An example of some devices that have IP Addresses are:

- Networked coffee machines
- IP-based lighting
- Networked TVs

You see, IP addresses are everywhere, and they touch almost everything these days.

So what is an address?

Well, an **IP address** is a number composed of 4 sets of numbers. These numbers can be assigned dynamically or statically which we will cover in just a bit.

A typical IP Address will look something like this:

```
Ethernet adapter Ethernet:

   Connection-specific DNS Suffix  . :
   Link-local IPv6 Address . . . . . : fe80::1142:7bed:752f:3cc%17
   IPv4 Address. . . . . . . . . . . : 192.168.0.10
   Subnet Mask . . . . . . . . . . . : 255.255.255.0
   Default Gateway . . . . . . . . . : 192.168.0.1
```

In figure x you can see that my IPv4 address is 192.168.0.10. Don't worry about what IPv4 means, just focus on the number.

When you look at my IP address you can that it has four sets of numbers separated by a (.).

The (.) represents both the network and host location of the device, I will cover networks and hosts in just a bit.

IP addresses can be public or private.

A **public IP** address is routable, meaning it can communicate outside its local network.

A **private IP** address is not routable, meaning it cannot communicate outside its local network. The reality is most of the IP addresses you deal with will be private IP addresses.

Static vs. dynamic addresses

Have you see the commercial for the Showtime Rotisserie?

If so you probably remember the old slogan "Set, it and forget it." Well, a static IP address is like that. All you need to do is set your **static IP address** on a device, and you're done.

There are a couple of problems with static IP addresses. For one, you have to be aware of what address you assigned to a device. In addition to this, IT groups are very cautious about giving static IP addresses to folks because they have limited IP addresses and need to monitor for address conflicts.

So what's an overworked IT person to do?

This is where dynamic IP addresses come into play. **Dynamic IP addresses** are addresses that are assigned to your device whenever it logs onto the network.

As you might expect the name dynamic, means the device is given an address from a pool of addresses (often called a subnet). These addresses are "leased" to the device for a period of time. After that period expires the device must "renew" its lease.

```
DHCP Enabled. . . . . . . . . . . : Yes
Autoconfiguration Enabled . . . . : Yes
Link-local IPv6 Address . . . . . : fe80::1142:7bed:752f:3cc%17(Preferred)
IPv4 Address. . . . . . . . . . . : 192.168.0.10(Preferred)
Subnet Mask . . . . . . . . . . . : 255.255.255.0
Lease Obtained. . . . . . . . . . : Wednesday, November 2, 2016 5:41:12 PM
Lease Expires . . . . . . . . . . : Monday, November 7, 2016 4:02:57 PM
```

The good news is dynamic IP addresses allow you to avoid the headache of managing addressing.

How do IP addresses get assigned

The way in which an IP address gets assigned depends on the whether the IP address is static or dynamic. If the address is static, then it will be assigned via manual entry. If the IP address is dynamic, it will be assigned to the device dynamically via a protocol called DHCP.

A full discussion of DHCP is way beyond the scope of this book, all you need to know is that DHCP dynamically assigns IP addresses to devices from a pool of addresses.

Where does this pool of addresses come from?

Well, my friends, this is where the concept of subnetting comes into play!

Sub-netting

Subnetting is to IT folks what PID loops are to controls technicians. Subnetting is a critical concept for IT professionals, yet a lot of folks misunderstands it. While I won't dive deep into the topic of subnetting I will give you the critical knowledge you need to be successful in your career.

In any given network you have a set number of IP addresses. This is not a network problem. It's simply because of how the current version of IP addressing (IP Version 4) was developed.

Because of this address shortage, IT professionals often need to divide up the IP addresses. A good way to envision this is to think of your radio. When you connect to the FM frequency, you are receiving one signal. Within this signal, there are many frequencies or channels.

The same is true with IP addressing. You can have multiple smaller networks inside a larger IP addressing range. Now I realize this may be vague to you. There is no real way to explain this without doing a technical deep dive which is beyond the scope of the book.

So instead of doing that, I'm going to ask you to trust me. Trust me that your IT group will need to know how many devices you have so they can slice (subnet) the right amount of IP addresses for you.

By the way, sub-netting can get rather complex when you have a large network. That is why it often takes IT so long to get you an IP address. They have to look through their network and find an address segment they can put your devices in.

Which brings us to the topic of networks!

What is a network

If I asked ten different BAS folks to tell me what a network was, I would probably get ten different answers. That seems like a cliché statement, but the reality is, that we as an industry have done a poor job of defining technical concepts.

When I do ask that question, though, I tend to get a few consistent themes that run through the answers.

Those themes are:

- Data flow
- Connectivity
- Access

Let's explore those.

According to Wikipedia a network is

"A telecommunications **network** which allows computers to exchange data. In computer **networks**, networked computing devices exchange data with each other using a data link. The connections between nodes are established using either cable media or wireless media."

From that definition, you can see the themes of data flow, connectivity, and access.

A network uses cabling or wireless transmission to provide a way to enable **data to flow** from one device to another. This **connectivity** is maintained through networking devices which control or regulate who and what can have **access** to the network.

A basic network is composed of 4 things:

- Host
- Switch
- Router
- Transmission media

The **host**, also known as a client, is a device used to access resources on the network. At the end of the day, networks exist to provide users with a way to access resources.

But what is a resource?

A **resource** is anything that a user needs to accomplish a task. Some common examples of resources are:

- Files
- E-mail servers
- Printers
- BAS supervisory devices

So how does the host access the resource?

On a local network, the host will connect to a switch via a switch port. A **switch port** is a physical port that allows the host to connect to the switch using an Ethernet cable, which is also known as the **transmission media**. There can be multiple types of transmission media depending on what you are connecting to the network.

I'm sure you've plugged a cable into the back of your cable modem before. Well when you did that, you connected your transmission media (Ethernet cable) to your cable modem's port.

A **switch** is a physical device, which allows multiple hosts on the local area network to connect to a centralized device. This device switches data between hosts on the local network. If you need to communicate to hosts outside your local network, you will need to route your messages through a device called a router.

A **router** is a physical device that uses routes to tell messages how to reach other networks. A **route** is a set of directions on how to access other networks, in order to communicate between networks.

The best way to demonstrate these concepts is to look at a couple of network examples.

Network example #1

In the figure below I have created a small network with two subnetworks, subnet A and subnet B.

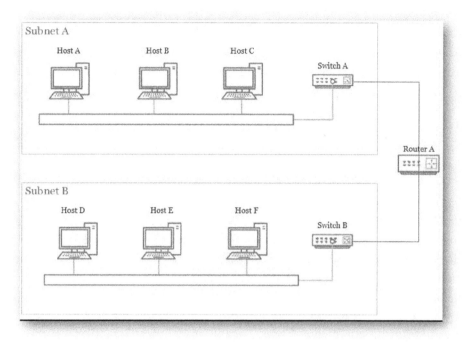

These subnetworks or **"subnets'** as you often hear them described are connected to a router, called Router A.

Router A connects the two subnets.

Each subnet has a switch that is connected to the router by a trunk link.

A **trunk link** exists to connect networking devices to one another. Remember we have networking devices and hosts (or end-devices).

Ok, so you have Switch A and Switch B connected to Router A via trunk links. You also have Hosts A-C and Hosts D-F connected via Access Links to the switch ports on Switch A and Switch B.

You still with me here?

I'm going to recap what I've covered so far:

- Networks have four key components hosts, switches, routers, and transmission media
- The hosts and switches are grouped into smaller networks called subnets
- In order, for subnets to communicate to one another they need to be told how to reach each other, this is called routing
- A router handles routing.

You ready to continue?

Awesome!

Moving right along. If Host A in Subnet A needs to communicate to Host B in Subnet A, then Host A will communicate through Subnet A's switch to Host B. This is illustrated below.

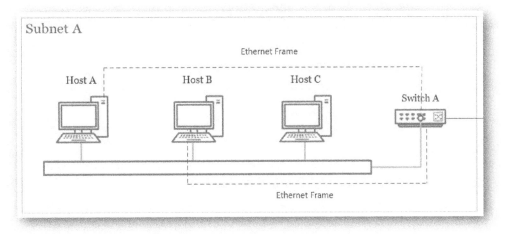

In the figure above you can see the message from Host A being sent to Switch A which then sends the message back out its switch port to Host B.

This exposes a key concept called switching.

Switching is the process of transferring messages from one network device to another network device on a local area network. The switch can do this because it keeps a list of which port connects to each device. By the way, just like we have trunk links we also have access links.

Access links connect the local devices to the network.

It is important for you to know what devices are involved in switching and routing. The reason you should know this is that there are different processes for troubleshooting switching and routing problems.

Let's look at a quick routing example.

Network example #2

Using the same network diagram, I am going to show you how Switch A can "route" a message (often called a packet) between Host A and Host E. If you want to communicate with another subnet that requires routing.

But what is routing?

Routing is the process of sending a message from one subnet to another subnet.

Now onto our routing example, in this case, we will be routing from Host A in Subnet A to Host E in Subnet B.

This example will involve the switches and Router A. This is an important example to pay attention to because so many of the network related service calls I have been on are due to routing issues.

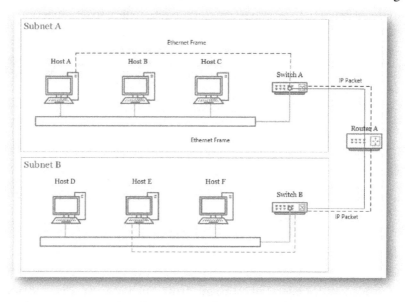

The figure above shows an example of a message being sent from Host A to Host E. In this example the host first sends its message to Subnet A's switch. Subnet A knows that Router A is its default gateway.

A **default gateway** is the router that all traffic headed outside of the local subnet is sent to. In this example, the default gateway for Subnet A and B is Router A.

Switch A will send the message to Router A via the trunk port between Switch A and Router A. Router A will look at the messages destination IP Address and will use its routing table to find out the "route" to other networks.

Router A will then send a message out its trunk port to Switch B. Switch B will then look at the destination IP address and locate the access port for Host E.

Networking is great and all but without servers to run our applications we'll get nowhere real slow.

What is a server?

Servers are the life-blood of your building automation system but sometimes they are improperly designed and improperly used.

If I had a nickel for every time, I went to a customer site and saw a server being used as the "front-end" I would have a lot of nickels.

Unfortunately, it takes a lot of nickels to make you rich, so I'm not quitting my day job anytime soon.

Back to the topic of servers. There are several questions I need to answer about servers, namely:

- What are Servers?
- What do servers do?
- Why should you care?

First off I'm going to discuss what a server does.

Have you ever accessed your company e-mail?

Sure you have, that's a pretty obvious question, right?

Well, in accessing your email you used a server.

You may be thinking, "Phil I used my laptop."

Yes, you did use your laptop, but your laptop was connected to a server that hosted the e-mail functionality. You used your e-mail client to connect to and run tasks on this server.

This is what is often called a client-server architecture, and it's why I strongly advocate against using your server as a front-end.

Client-server architecture exists so that multiple clients can connect to a server and run certain snippets of functionality. Before I break down the common server functions, let's get real clear on what clients are.

Clients are applications (these may be web browser based or physical software) that you use to access functions on your server. For example, if I want to view graphics for my building automation system. I will often open up Google Chrome and type in the name or IP address of my server.

My server will then provide me with a web page that allows me to view and edit my building automation system parameters. This functionality is called a web server, and it is one of the primary functions of BAS servers.

Are we good on clients?

Are you feeling comfortable?

Good news we are half-way through the chapter!

Moving right along, so I showed you what a client is, now I'm going to dig into the server.

So, what exactly is a server?

A **server** is a computer that is dedicated to performing a function for multiple devices.

As I discussed above, the client-server approach has the server responding to multiple client requests.

In the case of building automation systems, there are two server types.

Application server

The first server type is the application server.

The **application server** exists to provide access to the building automation or other smart systems software.

The end-user will connect to the server via a client device. The server will then provide this user with access to the building automation system. This allows the client device to avoid having to collect data from field controllers and other systems.

Database server

The second server type is the database server.

I will cover databases in the next section, right now just know that databases exist to hold data.

However, there is a lot to holding data. This is why it is common to have a dedicated server. The database server has a multitude of tasks:

- Storage of data
- Accessibility of data
- Modification of data

On a side note, the modification of data is often described by a concept known as the CRUD matrix. CRUD stands for create, read, update, and delete.

A database and application server will either be configured as a split-server package or a full server package.

Split-server

The **split server** package has the database and application servers on physically separate machines. This has the advantages of fault tolerance, lower computing requirements, and greater network speed. The disadvantages are that the owner now needs to run separate machines.

In my experience, once you get past ten buildings or 50,000 points it makes sense to split the application and database.

Full server

A **full server** has the database and the application server on the same physical machine.

This has the advantage of increased data access speed, lower cost of ownership, and easier maintenance. The disadvantages are that if the full server fails you lose both the application and database. In addition to this, your server can slow down if the BAS gets too large.

In my experience, a full server works for sites with less than ten buildings or 50,000 points.

What is a database

A **database** is a form of logical storage for data. This means that a database is a form of data storage with formatting applied to it. Folks often get the concept of databases and database management systems confused.

A database does nothing more than store data.

That's its sole purpose.

A **database management system** provides a way for you to create, read, update, and delete data (there's that CRUD matrix again!).

There are several different types of databases. The two I will cover are relational and non-relational databases.

Relational databases are the most common form of databases and are the database structure that most BAS's utilize. A relational database stores data based on relationships. The database will consist of a list of tables and keys. The keys will determine what data belongs to the tables. That is a very high-level explanation of databases, but it is all you need to know.

Non-relational databases sometimes use relationships but not always. This can be confusing for some people as the concept of semi-relational data doesn't always make sense. For example, you can have a flat row of data for a building and even though there is no relationship between the individual data, the data is formatted enough for you to know that all that data belongs to a specific building. Because of this, the data is semi-relational.

So why do you need to know this and why should you care?

The reason you should care is that many of the building owners I know are being asked to provide their management with increasingly detailed reports. If you do not understand how databases are structured, you will find it much more difficult to look at what data is available. If you don't know what data is available, how can you tell a database professional what data to extract?

Righting the remote connection

Remote connectivity is a hot topic right now.

On one side of the fence, you have folks who believe every building automation system should be isolated and that the BAS should never, ever, be on a production (business) network.

On the other side, you have people who believe that no-one wants to waste their time attacking BAS systems and that the security minded folks are just being paranoid.

My personal position lies somewhere in the middle.

Remote connectivity options

You have four options for setting up a remote connection to your building automation system.

Option 1: Public IP

In option one you give your BAS a public IP address, and you can expose that BAS to the world. In some cases, this may not be a bad choice.

For example, if you have a single office building and the BAS resides on its own network, then exposing the BAS to the internet is a low-risk scenario as long as you have a few safeguards in place (which I will discuss later).

Option 2: Virtual Network Connection

In option two you utilize a type of software called a Virtual Network Connection or VNC software. This is considered an application-to-application connection. The VNC software will create a one-time key that you can use to access your BAS server remotely. Some versions of this software allow you to have an account logged into to the server and the remote device. You can use this account to create VNC connections on demand.

Option 3: Virtual Private Network

In option three you utilize a Virtual Private Network. The difference between the VPN and VNC is subtle, but there is a difference. A VNC connection is a point-to-point connection, whereas the VPN is a point-to-network connection.

VPN is a point-to-network connection.

Did you notice the subtle difference?

With a VPN a tunnel is created between your system and the server. The VPN connection uses this tunnel to connect to a network. This network can have a single IP address and can still be isolated from the production (business) network, or, more commonly, this network can be a logically isolated virtual local area network (VLAN).

Option 4: A modem

Option 4 is an option that you don't very often anymore. On some systems, you can use a modem across a Plain Old Telephone (POT) line. This is an older approach, and while most modern control systems support POT's line connectivity, it is slow and less reliable.

Remote connectivity risks and how to counter-punch them!

So how do you mitigate the risk in these scenarios?

There are four risks that utilizing a remote connection to access your building automation system can expose you to. Those risks are:

- Someone breaking into your BAS
- Someone breaking into your network
- Someone breaking into server
- Someone breaking into network

Someone breaking into your BAS

If a bad person were to find your building automation system on the internet using a search engine they could attempt to break into your building automation system.

The first step to preventing this is to remove any and all default usernames and passwords. Default Usernames and passwords are easy to find on the internet through a Google Search.

Next, you will want to lock an account after a certain number of failed logins. This account should not be the same account you use to log in locally.

Finally, you should rotate your passwords every 90 days or when an employee leaves the business.

Someone breaking into your network

Someone breaking into your network becomes a potential issue when you are using a VPN or a server that has one of its network cards exposed to a public network.

In this case, you should be working with an IT group. With that being said, there are three tools you can implement if you have your BAS connected to a public network.

First off, you should implement an intrusion prevention and an intrusion detection system. While similar in name, an **intrusion prevention system** will block most attacks and an **intrusion detection system** will detect if attackers penetrate into your system.

Next, you should implement a firewall. A **firewall** will block specific traffic based on threat signatures. By the way, some would argue that a firewall is an intrusion prevention system.

Finally, you should prevent external browsing (meaning no ESPN or Weather.Com access on your BAS server). I know this is a pain, I hated having my server being blocked from going to Google because it made troubleshooting much more difficult. The reality is, there are a lot of potential threats on the web, and you just can't risk it these days!

Someone breaking into your server

The rules from the previous sections apply to your server as well. While I will go into much more detail in my book on BAS cyber security, the key point is to remember the acronym PPL (Patches, Ports, and Logins). Your system should be pretty secure if you keep your server patched (meaning the software is updated), open only the required ports (you may have to work with IT on this), and you limit what users can access the server.

Cyber security the fluff and the real

Risk comes from not knowing what you're doing. – Warren Buffet

There's a ton of bad advice out there and when it comes to cybersecurity. It's funny to me how folks who have never even done a security audit are advising people on how to secure their BAS!

Well, I'm going to help you end this, right here and now!

Key security concepts

I'm going to teach you what you need to know about cyber security. This won't be a lecture.

Nope, I'm going to give it to you straight based on the crazy things I've seen working in the smart building space.

But before I dive into this let's set one thing straight, security is contextual. If you run a two-story office building, you do not need the same security as the National Security Agency!

This fact seems to be lost on some folks who would make you believe that if you don't have a firewall, VPN, Intrusion Detection Software, and a dedicated security team that you are somehow doing something wrong.

You aren't, there it is, I said it...

But Phil, I don't need to or want to manage security!

Of course, you shouldn't be managing security. I totally agree with you!

Remember when I said earlier that security is contextual, well let's consider the context. If you are a sizable organization, chances are you have an Information Assurance Group, use them!

Wait a second! What about the folks without an IAG, are they just out of luck?

No, of course not, before I get into the nitty gritty of security I want to discuss how to determine when you should put cyber security controls into place.

Cybersecurity controls are not to be confused with building controls. A **control** in the cyber security world is a method or action that controls a risk. See in the cyber security space it's all about managing the risk!

Risk Assessment, Profiling, and Applying Controls

When I was pursuing my Master's degree in Cyber Security, the one thing they constantly pounded into our heads was that the goal of cyber security was to manage the risk associated with conducting business using IT resources.

Risk is the possibility of a bad guy attacking and damaging your business. The damage caused by a bad guy could show up in many different ways, but the ultimate goal is disruption.

When it comes to risk, there's only four things you can do:

1. Accept the risk
2. Ignore the Risk
3. Mitigate Risk
4. Avoid the Risk

Some would argue that accepting and ignoring the risk are the same thing. To make any decision on the risk, you need to understand what the risk is.

The way risk is analyzed is using a risk management framework. Now, to be clear, I don't expect you to start implementing risk frameworks and cyber security controls.

Rather, my intent in covering this topic is to expose you to these terms so that you can work with an internal IT group more effectively.

Once the risk has been properly identified the IT security group will work alongside the BAS owner to implement a series of controls.

Controls can be physical, administrative or technical. While an example of the three control types is beyond the scope of this book, what is important to understand, is that controls are used to reduce risk. One of the most common ways that controls are implemented is using a strategy called Defense-in-Depth.

Defense in Depth

First off, the days of isolated BAS's don't work. You've got an air gap to protect your BAS?

Ever heard of Stuxnet, it was transferred into a nuclear centrifuge on a thumb drive!

All of the security and isolation does nothing for you if I can physically access your machine. That's why there must be multiple layers of security. This multi-layer strategy is called **defense-in-depth**.

Defense-in-depth has often been compared to an onion. This is because a good security program should have multiple layers.

The idea is that you would look at the potential entry points an attacker would use to compromise your system. You then set up layers of security that get increasingly difficult for the attacker to peel back.

A tangible example of this would be physical security. If you consider a bank, it has multiple layers of security from the front door all the way to the vault door. Any one of these security controls could be compromised but when the controls are put together the chance of compromising them becomes increasing unlikely.

Chapter 4 quick summary

Yes!

Are you as excited as I am? I'm getting goosebumps as I write this.

Why you ask?

The reason I'm so excited is that I just exposed you to a world that is "voodoo" for a lot of folks. The most common struggle I hear from folks in the building space is around learning IT.

Now that you're through this chapter you have a solid IT knowledge base that you can use right now!

In this chapter, I also discussed how an IT group is organized and what roles each department serves. This will help you know who to work with on your BAS issues and let me tell you, knowing that information alone is worth the price of this book!

Next, I took you through most of the IT systems you will encounter during your daily adventures. I exposed you to networking, databases, and servers. You are well on your way to having the IT knowledge you need for a career in BAS.

Then you and I took a foray into the crazy world of remote connections. I talked you through how you can use remote connections you to access your BAS without being physically on-site.

I even showed you the four main forms of remote access which are:

- Public IP addresses
- VNC
- VPN
- Modems

How powerful is that? I mean seriously, folks ask me all the time, "Phil, how can I remotely access my BAS?"

Well, now you know, and **knowing is half the battle**. 10 points for anyone who can tell me where the quote I just bolded came from…

Finally, we closed out with a discussion on security. I showed you how security is all about reducing risk. I even taught you what risk is and how it can be dealt with through the use of security controls.

Ok, are you ready? In this next chapter, I'm about to make all of those hard to understand electrical concepts a whole lot easier…

CHAPTER 5

ELECTRICAL FUNDAMENTALS AND ENERGY MANAGEMENT

What's in this chapter?

In this chapter, I'm going to take you through the fundamentals of electricity and energy management. Before I do this, I want to level set on what the goal of this chapter is. The goal of this chapter is not to make you an electrician or an energy management expert.

The goal of this chapter is to teach you the fundamental information that you need to know to be in the building automation profession.

That's why in this chapter, I am going to be staying at a high-level rather than going deep into the topics I am covering.

So what topics am I covering you ask?

Chapter 5 will cover the following topics:

- Electrical fundamentals
  - Ohm's law
  - What electricity is
- Energy management
  - Measurement verification

Will you be able to go start running wiring, connecting transformers and performing measurement verification after this chapter?

No.

Will you understand how electricity and energy management are related to building automation systems?

Definitely, and in my opinion that understanding is something that is missing in our industry.

Electrical fundamentals

Man, oh man, what a big topic to unpack. When I started to think about how I was going to write up this chapter, I thought about the things I wish I known back when I started my building automation journey.

Throughout my career, there were some specific topics that kept rearing their ugly head. As I look back, I realize that if I had had a solid understanding of these topics, I would've avoided a lot of pain, been better at designing and installing building automation systems, and been able to interact with my electrical subcontractors more efficiently.

That's why in this section I'm going to discuss some key electrical concepts. I will tell you, though, just because these are concepts does not mean they lack importance.

Quite the opposite.

I can promise you one thing. If you get these concepts down, everything else you do from an electrical perspective, involving building automation systems will be easier.

One quick note, I know I have a diverse set of readers who are going through this book with me. If you are already comfortable with the topics I'm covering, then, by all means, feel free to skim through this chapter.

However, as I said, in the beginning, the book I still encourage you to at least skim through. Who knows, you may be reminded of something you forgot...

One more thing, some of what I'm about to describe may seem a little hard to understand for some of you. That's okay; it took me a while to figure all this stuff out.

If you simply gain a familiarity with these terms and the relationships between these terms, I will consider this electrical fundamentals section a success.

Ohms law

Ohms law often expressed as Current = Voltage / Resistance, or I = V/R, describes the relationship between current, voltage and resistance.

This is the most important electrical formula you will learn.

Understanding the relationships that exist between current, voltage and resistance, will help you to understand how the electrical systems, like VFD's work.

The cool thing about Ohms law is that if you know two of the three variables (current, voltage and resistance) you can determine the third.

I will demonstrate how this works as I go through each of the three variables.

Current (Amps)

I= V / R

Current is the flow of electricity, technically electrons, across a wire. As electricity flows across a wire, the resistance and voltage affect the amount of current.

Current is measured in amps.

Voltage (Volts)

V= R x I

Voltage is the potential energy that could be released across a wire. Notice that I said the potential energy. As I described in the current section, when voltage is released across a wire encounters resistance, or impedance, that in turn creates current.

Voltage is measured in volts.

Resistance (Ohms)

R = V / I

The last variable is resistance. **Resistance** indicates the difficulty of passing electrical current through a wire. As electricity flows across a wire it encounters certain levels of resistance, the amount of resistance will determine how much electricity flows through the wire, also known as current.

Resistance is measured in ohms.

Power, volt amps and power factors

Now that you understand how different aspects of electricity are measured, I'm going to explain to you how electricity is used to get things done.

Power (Watts)

I'm sure you have encountered power multiple times throughout your life, although you may have never heard "power" called power. Let me introduce you another name that power goes by.

Power is also known as **wattage,** and naturally, it is measured in **watts**.

Without going too far down the rabbit hole, it's important to understand that power is a measurement of electrical energy and is known as real power.

Volt-amps (VA)

If you've worked in the building automation space for any amount of time you've probably heard of the term volt amps. Often the power requirements for actuators and building automation controllers are measured in volt-amps.

But have you ever wondered what a volt amp is?

A **volt amp** is a measurement of the voltage being applied to a device multiplied by the current that the device is drawing. Volt-amps is also known as apparent power.

Power factor

You may have heard the term **power factor** used to describe the efficiency of a circuit. Often, you will hear folks saying that they want to increase their power factor.

Why would they want to do this?

The reason folks want to increase their power factor is because power factor is the ratio of active power to apparent power.

Okay, that's a mouthful.

Here's what that means let's say you have 100 W of real power, and your circuit has a power factor of 60%.

This means that the devices at the end of the circuit will only receive 60W of power. This will lead to some significant inefficiencies especially when you're already paying for a certain amount wattage from the utility provider.

Your utility provider promises a certain amount of wattage to you. It is up to you to effectively distribute this wattage throughout your building.

Electricians can use a device called a capacitor to increase a circuit's power factor.

That's about as deep as I am going to go on that topic. If you would like to know more, I would encourage you to look into some of the programs offered by the Association of Energy Engineers (AEE)

Specifically, their five-day Certified Energy Manager (CEM) boot camp. I view that course as being one of the more useful courses I have paid for in my building automation career.

So, what is electricity?

At this point, you should have a solid understanding of electrical fundamentals. Now, I am going to discuss alternating-current and direct-current, also known as AC\DC. Join me in the next section if you're ready to rock!

Because building automation systems use both alternating-current and direct-current it is important for you to understand what they are and their differences.

Back in high school, you were probably taught a concept called the sine wave. The **sine wave** is a curved line that oscillates, or goes up and down, from a zero point. This wave is also known as the waveform.

In the world of electricity, there are two forms in which current is transferred.

Can you guess what those are?

If you guessed alternating current and direct current, then you are correct!

Alternating current

Alternating current is a form of power that spans across the entire waveform. This means that it has both a positive and a negative charge is also known as polarity. This is why you can connect the hot and common wires on either terminal of your building automation controller.

Direct current

Direct current is another form of power. DC power only transfers across a specific pole of the waveform.

What exactly does that mean?

What that means is that DC power can have negative or positive polarity, but it cannot have both. If you connect the wires from a DC sensor to the wrong terminals on a building automation controller, you will get a negative value instead of a positive value.

Energy management

Energy management, is the process of monitoring, controlling, and conserving energy. I could write a whole book on how energy management is "done."

In this section, I will familiarize you with the main functions of energy management that I've seen building automation professionals involved with.

How to monitor, control, and conserve energy

Your building automation system can help you monitor, control and conserve energy. I will discuss how you accomplish these functions in this section.

There are three primary energy management tasks that a building automation system can be utilized for. These tasks are metering, measurement and verification, and load-shedding. There is a specific reason why I put these three tasks in the order I did.

As I see it, you need to perform **metering** first to gather a baseline. Once you have this baseline, you can use the **measurement and verification** process to compare the baseline to the actual energy usage. Once you know how far you are from your baseline, you can implement **load shedding** to reduce your energy loads during peak periods.

Metering

A building automation system can be used to meter "utilities." Typically, this is done by either integrating an IP-based meter into the building automation system or hardwiring a meter into a building automation controller.

This meter data is captured and stored in a trend database. You can access your data from this database at a later point in time.

I covered meters in chapter 3.

Measurement and Verification

Measurement and verification is the process of measuring the use of energy within your buildings and systems.

There are several ways you can perform measurement and verification. The method I prefer to use is called the International Performance Measurement and Verification Protocol or IPMVP. The IPMVP comes from the Efficiency Valuation Organization.

The **IPVMP** is a series of books, three to be exact that covers energy and water efficiency, demand and load management, and renewable energy. I will be focusing on Volume 1, which includes is the measurement and verification processes

Volume 1 provides an introduction to measurement and verification, as well as an approach, measurement, and verification plan, and series of "methods" for performing measurement verification.

I will not be diving into the specific M&V methods in this book. Rather, what I want you to take from this section is that to perform measurement and verification, you need to have a solid metering system and utility baseline.

Load-Shedding

There are hundreds of potential energy saving retrofits and upgrades that can be implemented with a building automation system.

With so many potential ways to save energy, why did I pick load shedding out of all the different strategies that could be enacted with a building automation system?

The reason I picked load shedding is that utility providers are increasingly requiring or rewarding organizations that implement load shedding.

Load shedding is the act of purposely reducing the utility consumption of your building by either changing the way your systems operate or by turning off your systems.

There are three levels of load shedding:

- Level I involves the user manually choosing to adjust each systems performance
- Level II involves the user manually choosing the systems to turn
- Level III is where the utility has the capability to tell the users building automation system what amount of power the building should be consuming, and then the building automation system will automatically shut systems down

There's a lot of controversy around load-shedding, especially in the United States.

This seems to stem from folks not wanting to give up the control of their building to regulatory bodies.

Chapter 5 Quick Summary

Oh, snap! Or maybe spark…

You just learned the fundamentals of electrical systems and energy management. Wowzers!

Now, don't let this knowledge go to your head, I don't need you trying to install transfer bars and switchgear…

In this chapter, I went through the fundamentals of electricity. I taught you Ohm's law and revealed how electricity works!

Next, I went into energy management. I talked you through energy management, measurement and verification and load shedding.

Think of the things you can do with this knowledge! Speaking of knowledge in the next chapter I'm going to expose you to the world of professional organizations and standards…

CHAPTER 6

STANDARDS AND ORGANIZATIONS

What's in this chapter?

Standards and organizations, where do I even begin?

This chapter could go in a million different directions. Because of that, I am going to focus on the most common standards and organizations that you will encounter during your building automation career.

Will I miss a few organizations or standards that you feel should be in here?

Probably.

Will I cover most of the organizations and standards you need to know about?

Definitely.

In this chapter I will be covering:

- Professional Organizations
  - ASHRAE
  - BOMA
  - CABA
  - USGBC
- Standards
  - ASHRAE 55
  - ASHRAE 62.1 and 62.2
  - ASHRAE 90.1

For some of you, this will be knowledge you already have, for others, this will be the first time you have heard about these standards. Either way, I look forward to covering these topics with you.

Professional organizations

There are several professional organizations out there that focus on buildings, but only a few of these organizations also focus on building automation systems. In this section, I will discuss these organizations.

ASHRAE

The American Society of Heating, Refrigerating, and Air-Conditioning Engineers (ASHRAE) was founded in 1894. ASHRAE is the largest professional organization for engineers who focus on building systems. ASHRAE is massive, it has chapters (that's what ASHRAE calls its local groups), across the globe.

In the building automation world, ASHRAE is best known for ASHRAE Standard 135, which is the BACnet standard. As you know from Chapter 2, BACnet is one of the most popular building automation protocols in the world.

ASHRAE is also known for several other standards that it has created. Several of these standards apply to building automation, and I will cover those later in this chapter.

You can access ASHRAE at www.ashrae.org

BOMA

BOMA stands for the Building Owners and Managers Association. BOMA was founded in 1907 and is focused on a variety of topics that are important to building owners and managers.

BOMA is a good resource if you're looking at increasing the energy efficiency and sustainability of your building.

You can learn more about BOMA by going to www.boma.org

CABA

The Continental Automated Building Association, also known as CABA, is focused on advancing intelligent home and building technologies.

Now what exactly does advancing building technologies mean?

As I was looking through the meeting minutes of the CABA Intelligent Building Council, I was able to pull out a couple of key initiatives that they are working on.

These are:

- Intelligent buildings and cyber security research
- Improving organizational productivity with building automation systems
- Intelligent buildings IoT roadmap

Several building automation companies support CABA and CABA is one of the only organizations that solely focuses on building automation.

You can learn more about CABA by visiting www.caba.org

USGBC

USGBC stands for the U.S. Green Building Council. The USGBC is known for its creation of the LEED program.

You can find out more about the USGBC by going to www.usgbc.org

Standards

There are a ton of standards that are related to the building automation space. However, there are three main standards that pop up on almost every project. Understanding these standards will affect how you design, manage and service a building automation system.

These standards are:

- ASHRAE 55- 2013 Thermal Environmental Conditions for Human Occupancy
- ASHRAE 62.1-2016 Ventilation for Acceptable Indoor Air Quality
- ASHRAE 90.1-2013 -- Energy Standard for Buildings Except Low-Rise Residential Buildings

ASHRAE 55-2013 Thermal Environmental Conditions for Human Occupancy

As I mentioned in chapter 1, ASHRAE 55 defines the comfort zone. This concept comes out of the 55-2013 standard. This standard suggests the set point for the building based on six factors. These factors help the engineer determine the correct setpoints for addressing the thermal comfort of building occupants.

The six factors that this standard looks at are:

1. Temperature
2. Thermal radiation
3. Humidity
4. Airspeed
5. Personal activity level
6. Personal clothing

This standard becomes especially important if you are operating a building outside the standard 72° setpoint.

ASHRAE 62.1-2016 Ventilation for Acceptable Indoor Air Quality

ASHRAE 62.1 is focused on defining the minimum ventilation rates based on space types. You may be asking, the minimum ventilation of what?

This standard defines the amount of outdoor air that needs to be provided to a space based on its projected occupancy levels.

For example, a college lecture hall would need to provide 7.5 CFM of outdoor air per person. Therefore, if you had a lecture hall with a capacity of 200 people, you would need to provide 1500 CFM as your minimum outdoor air.

Imagine you are in a geographic area that is hot and humid. You now have to condition 1500 CFM of hot, humid air.

This is why understanding standards are so important. Since building automation technology now exists that is capable of counting how many occupants are in a space you can reset the minimum outdoor air set point. Naturally, this could result in huge energy settings.

Another way to control the minimum outdoor air set point is a CO_2 sensor.

ASHRAE 90.1-2013 -- Energy Standard for Buildings Except Low-Rise Residential Buildings

ASHRAE 90.1 has been a popular standard lately because several green rating programs like LEED are using it as the standard of choice for establishing the level of energy efficiency that a building should reach.

The standard dictates out two paths for complying with the standard.

Prescriptive path

The first path, called the **prescriptive path**, focuses on defining the minimum requirements for the following parts of the building:

- Building envelope
- Domestic water
- HVAC
- Lighting
- Other equipment
- Power

This path relies on setting minimum requirements for compliance. A designer or building owner can utilize the building automation system to implement sequences that increase the efficiency of these systems.

An example of this would be using a VFD to increase HVAC efficiency or automating the use of daylighting to reduce lighting loads.

Performance path

The second path, called the **performance path**, utilizes a baseline design energy model, often called an **energy cost budget** or ECB. This ECB is created using modeling software and is very similar to Option C of the IPVMP.

The goal of the second path is to have the proposed design model be less than or equal to the baseline design model.

Now think about that for a second, the efficiency of the building is determined by comparing the proposed model to the baseline model.

Does that set off any red flags?

As any experienced building designer or operator knows, a building will have a completely different energy model 1 year after the certificate of occupancy.

Why is this?

This decrease in energy efficiency often occurs because systems will be put in hand, set points will be changed, and electrical/heat loads will be introduced to spaces.

My thoughts on the two paths

If you simply want to get credit for complying with the standard or you are looking for LEED compliance, then you should pursue the performance path. If you want to ensure that your systems are efficient, then you should pursue the prescriptive path along with a solid measurement and verification strategy.

Chapter 6 quick summary

In this chapter, I showed you the professional organizations and top three BAS standards that you should know about. And on that note, we're done with section 1!

Oh my goodness! Can you believe it?

You just gained a level of knowledge that already places you heads and shoulders above most of "the other folks" in the building industry.

You can walk around with your shoulders back, and your chin raised high knowing that you now know more than most of the folks out there!

Well, my friends, it's time to grab a coffee and strap yourself because section 2 is going to take your knowledge to the next level!

Section II – The Art of Buying, Installing and Upgrading a Bas

Section Overview

Ok, now that you have the basics down I am going to ramp it up a notch. In this section, I am going to discuss the topics of buying, installing and upgrading a BAS. Without these steps, nothing happens. You can have all the BAS knowledge in the world but without a BAS that knowledge is useless.

In these chapters I am going to walk you through the process of buying, installing and upgrading a BAS. By the time you finish this section, you will understand how the processes for purchasing, installing and upgrading a BAS work.

CHAPTER 7

PROCUREMENT: THE ART OF BUYING STUFF

What's in this Chapter?

It's time for everyone's favorite topic money!

At the end of the day none of the stuff I'm teaching you matters if you can't buy any of the building automation parts and pieces I've been discussing up to this point.

How do you buy building automation systems?

How do you ensure you're getting the best value (notice I didn't say price)?

How do you know that what you're asking for is what you're getting?

These are challenges that owners, contractors, and engineers are facing every day. How you approach the procurement process can make the difference between you getting a antiquated geriatric time clock or robust kick down the door take no prisoners building automation system.

That may seem a little overly dramatic for some of you, but the reality is every day I see specifications and designs come across my desk that I know are not going to meet what the customer is looking for.

In this chapter, I'm going to take you through the procurement process, and I'm going to show you exactly how a product is procured. By the end of this chapter you will understand:

- The difference between private and public procurement
- The common procurement process
- The differences between RFI, RFQ, And RFPS
- Pricing methods and strategies

If you're ready to learn how to make sure that the procurement process is as painless as possible, then please join me in the next section.

What is procurement?

Funny enough, this was the 2nd to the last chapter I wrote. I went back and forth on how to write this chapter.

And if I'm honest with you, at one point I even considered removing this chapter from the book.

Why you ask?

Well, it wasn't because procurement is unimportant. It also wasn't because there's was a lack of information about procurement.

Nope, it was because I don't have expertise in procurement.

Now, before you shut down and dismiss this chapter, let me explain.

Sure, I've bought tons of "stuff" while working at a large Fortune 100 building automation company. I also managed the partner ecosystem for a Fortune 100 company where I oversaw the negotiation of terms, conditions, and pricing for massive supplier agreements.

However, I've never been a procurement person. When it comes to contracting, integration, upgrades, service, you name it. I've got the real-world experience.

Procurement, not so much…

So, how could I resolve this dilemma and give you some real world education on this critically important topic?

Elementary my dear Watson.

For this chapter, I went and interviewed procurement professionals in both the public and private sectors and asked them a series of questions. What follows is a summary of the information they provided me.

Private vs. public procurement

Our journey begins by discussing the differences between private and public procurement.

The reason I start here is that there are significant differences between how you procure equipment and services depending on the procurement process you use.

As it was explained to me, procurement processes break out into two buckets. Those buckets are public procurement processes and private procurement processes.

Before I go into describing public and private procurement, I wanted to clarify that the public and private classification has nothing to do with whether the company is a public or private company. Private stands for a commercial organization and public stands for a government organization.

Public procurement

The public procurement process is a highly regulated process. There are procurement laws and regulations that must be complied with. Ultimately the public procurement process is intended to foster transparency and efficiency.

Now I know some of you out there chuckling because at times it seems that the public procurement process is anything but efficient. The intent of the public procurement process is to get the best price for a product or service anytime taxpayer money is spent.

Realize that when I am talking about public procurement, I am focused on the United States. I do not claim to have experience or expertise around how the public procurement is handled outside the United States, although I would imagine it is somewhat similar.

Public procurement does make it a little bit difficult for both the buyer and the supplier if an existing relationship has not been established or the buyer wants to use a specific vendor. This can be resolved but often it can be quite difficult.

This is where standardization comes into play.

One of the main methods I have seen used to ensure that a specific vendor selected is to form a standard that has been ratified and agreed upon. The standard can then be used to procure material from a specific vendor, depending on the local procurement laws.

The nice thing about the public procurement process is that it tends to follow a structured approach that consists of RFI's, RFQ's, and RFP's (I will describe what each of these terms mean a little bit later in the chapter).

The downside of public procurement is that price often drives the procurement process. This can sometimes result in substandard design, products, or services.

Private procurement

If the public procurement process is a structured organized process, then the private procurement process is the Wild West. In the private procurement process, there are significantly fewer laws and regulations that are imposed upon private companies.

Because of this the method for reviewing, selecting and procuring products and services, is often dictated by the company.

This provides the benefit of a flexible process, in which the most valuable product or service can be selected. Unfortunately, this also opens the door for "buddy relationships," where personal relationships can cloud the procurement process. The resulting solutions that are not selected based on price or value but rather the relationship the customer has with the salesperson.

Don't get me wrong. I'm all for forming a relationship with your customer or buying from people you like. However, when you let that relationship influence you to the point where you're buying a clearly substandard product, then you are not doing your due diligence as a representative of your company.

Now that may seem obvious to you, but you would be shocked how many times I've seen no name products brought onto a project because of a relationship based sale.

The downside of private procurement is that the procurement process can be different for every company.

Because of this, the private procurement process can be quite ineffective at times.

Common procurement processes

When someone is beginning to look for a building automation system, the process tends to flow from information gathering to qualifying, to purchasing. In the building automation world, heck, in the overall construction world we tend to call these processes request for information, request for qualification, and request for proposal.

Initially, the customer will start with a vision and a set of outcomes. Based on these outcomes further information will be requested. This information will lead to a set of questions that are used to qualify potential products. Finally, the product manufacturers will provide a proposal for the potential project.

In the following sections, I am going to talk through each one of the steps in greater detail.

Request for Information (RFI)

So you want some information huh?

Pay close attention to this step folks, because so many people skip it!

The Request for Information (RFI) is the first step during the procurement process, or at least it should be. In this step, the customer, sometimes with the assistance of a design engineer or consultant, will create a design narrative that will be sent out to a list of contractors.

Typically, this information is sent out with a request for information on how this design narrative can be accomplished. The result is that the contractors will respond with their interpretation of the narrative and how they believe they could accomplish the narrative.

Based on these responses the owner and their representative will end up creating a functional specification. By the way, this is why so many contractors want to get in front of the RFI process.

You see if a contractor's RFI response becomes the functional specification, often known as the basis of design, then the contractor will have a higher likelihood of being selected later when the project goes out for bid.

Whether it's arrogance, ignorance, or simply a perceived lack of time, I've seen so many projects not meet the intent of the customer because the owner never asked for the information that could have defined their intent.

So how do you ask for this information?

How do you go and properly request information?

Furthermore, what information do you even ask for?

These are all questions I cover in my audio course titled *How to Evaluate a Building Automation System*.

Request for Qualification (RFQ)

The Request for Qualification (RFQ), is the second step that is taken during the procurement process. This section is going to describe how the RFQ process works.

Once the Request for Information (RFI) has been responded to, the customer or the customer's representative will collect these responses. These responses are then used to create functional specifications that detail out the expectations of the customer. These specifications typically will not get into the technical details of the solution. Rather they will detail out how the solution should function.

Now that the functional specification has been created it will be sent back out to a list of contractors who will have to meet a set of qualifications based on the functional specification.

Based on my experience these qualifications tend to be:

- Communication protocol
- Control system architecture
- Procurement availability
- Local presence
- Experience performing a similar functional specification

Each contractor will respond to this functional specification, which usually includes a qualification checklist that the contractor needs to fill out. Upon completing the RFQ, the submissions will be reviewed, and the contractors that qualify will be invited to respond to the Request for Proposal (RFP).

Request for Proposal (RFP)

The Request for Proposal (RFP) process is typically the final step in the procurement process. In this step, one of two things will happen. Either the customer or the customer's representative will have formed a detailed specification that contractors will be responding to or the contractors will be responding to the functional specifications with a detailed proposal.

Typically, the proposals that contractors submit during this phase will be considered the final price.

There are quite a few nuances to the RFP process that you can only pick up with experience. These nuances are things like:

- Final cost versus maximum price
- Qualified responses versus responses
- Total cost versus block hours

Final cost versus maximum price

Depending on how the RFP is worded, a contractor can respond with a final cost proposal or a maximum price proposal, sometimes known as a not to exceed proposal.

The reason a contractor may want to respond with a maximum price is that the functional specification or scope may be missing key details that can only be discovered once the project has been executed. Obviously choosing a maximum price in lieu of a final cost proposal is in favor of the contractor.

Unfortunately, some sites have been kept in such disrepair that there's no way a contractor could be reasonably expected to fully account for all costs associated with the scope of the project.

Which brings us to....

Qualified responses versus responses

Sometimes if allowed contractors will qualify the responses based on their interpretations of the specification or scope. Contractors will also qualify their responses based on site conditions. This is much more common on retrofit projects than new construction projects.

Some specifications and scopes are unclear, and some project sites have not been maintained. When a site hasn't been maintained the only way an owner can get contractors to bid a job is to accept qualified responses.

Some of the scope items that are usually qualified are:

- Controller or communication trunk capacity
- Control device conditions (meaning are things falling apart or do they still work)
- Accessibility to the control system

Total cost versus block hours

Sometimes instead of responding to a scope with a flat price a contractor will choose to propose their price as a block of hours.

Block hours pricing is where the contractor estimates how many hours it would take to perform a task and then allocates a block of hours to that task. This form of pricing is often associated with maximum price or not to exceed pricing.

Bid forms and segmenting costs

A lot of companies do just fine with the procurement process until it comes to collecting the bid responses and segmenting the costs. In the construction, world contractors have become rather effective at handling bids and segmenting costs. The problem is that companies typically do not perform the same amount of large-scale projects as contractors.

So then how does a company go and gain the knowledge of how to handle bid forms and segment?

This becomes an even harder problem to solve when you consider that contractors are often forced to value engineer their prices. Because of this, a contractor is very cautious when it comes to segmenting out their prices.

I mean think about it. If you were to ask a contractor who had to "be low" to win a project to segment their costs any further, how would they do that?

I mean seriously, that contractor would have no cost left to handle any risk in the project. I'm sure some of you are saying "that's not the customer's problem." The fact is contractors are asked to bid projects without ever being given all of the information and in some cases, they aren't even allowed to visit the site.

How is that reasonable?

When contractors want to cover their risk, and owners want all of the cost exposed, who wins? I don't claim to have the answer for this, after all, folks a lot smarter than myself have been trying to solve this for a long time.

However, what I can do is show you what I believe to be an effective process for using bid forms. I will also show you a cost segmentation approach that I believe is fair to both the owner and contractor.

Bid forms

In my opinion only using bid forms to evaluate quotes from your contractors is a practice that at the least is ineffective and at the most is just plain idiotic.

Now, why do I say that?

Who am I to go and say that a practice that has been used for decades now is a bad practice?

The reason why I say using only bid forms is bad practice is that it only gives you a part of the picture.

You see, bid forms are typically segmented to break out the costs associated with specific tasks. As you will learn later in this section, you can take three different contractors and get a different approach to the same scope of work.

If you don't evaluate contractors the approach, then you may pick a contractor who is going to give you a substandard system, if the system even works at all.

What I propose is a modified version of the bid form. While you should still have a bid form, it should include the following extra categories:

- Experience performing this type of project (low, medium, or high)
- Number of staff required to man the project
- Current staff levels
- Current number of other projects
- If a specific expertise is required for the project (for example coding or graphics)
  - Find out how many people with these skills are working for the contractor

Those may seem like some weird things to ask for on a bid form. However, remember we are talking solely about building automation systems in this book. It is my experience that projects tend to go wrong because of the following things:

- Improper scope definition
- Contractor's experience with this task
- Contractor's inability to "man the job."

There's one more area that I need to cover, the two words that make every contractor cringe and those are, cost segmentation.

Cost Segmentation

Cost segmentation sucks, there's no if, and, or, buts about it.

I'm going to ask you to take a second and think about this from the contractor's point of view. So here you are, a building automation system contractor, and you are asked to bid a project.

Naturally, you go and ask for the scope or the specifications. You're told, the specifications are partially done, and you will have to look at the old drawings and figure out how you would approach the job and bid it accordingly.

Okay, not ideal, but you've done this before, and you go and count the pieces of equipment. You can go and quickly come up with a general range of how much it will cost to put controls on those pieces of equipment.

Then you go and add about 5% to 10% extra on top of the project because of the risk.

Next thing you know, you're shortlisted (meaning you are one of the companies that will move forward in the procurement process) and you're asked to break out your costs.

Crap!

How can you break out your costs, you don't even know for sure what's on the job?

But, it's about to get worse. Next, you are sent a bid form, with a cost segmentation section, and you are told to break out your costs for each section.

Okay that's fine, you do that, and knowing that the biggest risk is the labor portion, you shift a significant portion of your costs over to the labor bucket on the bid form.

A week or two goes by, and you don't hear anything. You reach out to the customer or contractor that you are bidding to, and you ask what's up?

It's now that you're told that the contractor or customer decided to go with your competitor because their labor was lower than yours.

Now, I know what some of your thinking, the building automation contractor just needs to get more efficient and their labor costs won't be so high.

But what that means is that the building automation contractor needs to absorb the risk of not having a clear scope. Now later in the book, I will show you exactly how to have a clear scope, so this won't be an issue on your projects.

However, if you are experiencing this right now you probably want to know how to take care of it, because honestly while it sucks for the building automation contractor, it sucks even worse for the end customer because they end up getting a low-bid surprise.

How can you segment cost and not end up in the situation?

I've got three ways that will help you avoid this situation, but before I go there, I've got one more story.

I was working with a local university on a project. I had a great relationship with the university, and they wanted to use my company for the project, I mean after all I worked there and who doesn't want to be on a project with me (okay don't answer that truthfully).

Like I said, I was working with a local university, and one of our competitors came in at $30,000, and we were at $100,000. Obviously, there was quite a disparity in price. At this point, there was nothing that the university could do to justify a $70,000 cost difference.

Now the university knew that our competitors were purposely coming in low so that they could get the job. However, no one comes in $70,000 low.

Long story short I was able to use one of the three methods I'm about to teach you, to win the project.

Method one: Apples to apples pricing

The first thing I did was I met with the university and level set around the scope. Now this is where you have to have a relationship with the owner or with your building automation contractor.

You see I went to the owner and said,

"Hey, there's no way these folks can be $70,000 then us."

I then asked if the owner could describe the other contractor's scope of work to me without giving me the price, name of the contractor, or any of their product numbers.

Once I heard the proposed scope, I realized very quickly that this contractor was using electro-mechanical wall mount thermostats with BACnet\IP connections instead of building automation controls.

Now technically all the scope of the project said was that the BAS needed to control the fan coil units in the building.

However, I knew that this customer wanted to have DDC controllers mounted on the fan coils. After going through my scope and the other contractor scope, I identified the differences and pointed them out to my customer. I instructed the customer that if he wanted me to match the competitor's scope I could do that.

I hope you all noticed what I just said. Instead of looking at the two different scopes, all that was looked at was the price. The customer was about to buy the same kind of thermostat I have at my house for a university building.

Method two: Validate external costs
I've seen several projects where costs external to the contractor were purposely excluded from the building automation contractor's price. A couple of examples of costs that I've seen excluded are:

- Software licensing
- Third-party integration costs

- External contractors
- IT costs

Unfortunately for the ethical building automation contractor these prices will typically get put in the material portion of the cost breakout. That is why for all BAS contractors I recommend that you write in a section for external costs, on all cost segmentation models or forms.

I know from experience that if you do not clearly call out that external costs must be exposed you will end up receiving a change order during the project to handle these costs, usually after the project has been completed.

Method three: Look at the skills required

If you recall in the bid form section, I recommended that you consider the skills that are required for the project. If a project requires special skills, like database, network, or programming skills, you should include a section for these.

Chapter 7 Quick Summary

I can hear the gears turning in your mind. For some of you, this is the first time you've been exposed to this information. Well, my friends, it's about darn time :-D

In this chapter, I took you on a magical journey through the crazy world of procurement. I showed you the differences between private and public procurement along with the common procurement practices that are used.

Next, I moved through the RFI, RFQ, and RFP process. I showed you how this stuff works in the real world.

Finally, I took you through an analysis of the different pricing methods and ways to evaluate proposals. I even told you a little story about a time I had to defend my castle against an unscrupulous contractor…

This next chapter is going to answer the age old question.

How are buildings built?

CHAPTER 8

How Buildings are Built: A Look at the Construction Process

What's in this chapter?

The construction industry is an interesting industry.

For most people, the construction process is one of these things that they only go through one time in their career. In this chapter, I am going to share with you the lessons I've learned working on over 250 major construction projects.

As I take you through this chapter I am going to discuss the following topics:

- How is a project executed
- Delivery and funding models for a construction event
- Selecting a tier 3 contractor
- Project delivery models
- Design phases
- The project process

How is a project executed

I think the best way to teach you about the construction process is to walk you through how a construction project takes place.

A construction project starts when an architect, who's often called a **master planner**, meets with an owner, and together they decide on a concept for a building or set of buildings. It is this concept that will eventually become a building.

The owner will work with the architect, as part of this process, to decide on what he or she wants to achieve. This phase of the construction process is usually focused on answering the following questions:

- What experience does the owner want the building occupants to have?
- How does the owner want the building to operate?
- What is important for the building to have?

Based on this information the architect will create a conceptual design for the building. To be clear, this is not a construction design. The owner and architect will use this conceptual design to come up with a conceptual budget.

Selecting the general contractor

It's at this point that the owner and the architect work together to select a general contractor.

Some of you may be asking, what is a general contractor?

A **general contractor** is responsible for overseeing all of the subcontractors who are working on the project.

Which brings me to the next obvious question.

What is a subcontractor?

A **subcontractor** is someone who works under a contractor. On a construction project, there are multiple tiers of subcontractors. These are typically called tier 1, tier 2 and tier 3 contractors.

A typical project will be organized with a general contractor and architect in the **tier 1** role. Tier 1 contractors are also known as the "owner's team."

Tier 2 contractors are usually made up of contractors like the technology, mechanical, electrical, and plumbing contractors. The mechanical, electrical, and plumbing contractors are often known by the acronym (MEP).

Each of the second tier contractor's will work on selecting their specialty subcontractors or Tier 3 contractors. In most cases, the building automation contractor is a tier 3 contractor and will be hired by the mechanical contractor.

I will go deeper into the selection process of Tier 3 contractors, but before I do that, I want to take you through some of the delivery and funding models that are used for a construction event.

Delivery and funding models for a construction event

As I said the first contractor, the owner will select is the general contractor. The general contractor will work with the owner to decide on a funding vehicle for the project.

There're a couple of different funding vehicles. The ones I will cover are the main ones you will see in the marketplace as of 2016. These funding vehicles are:

- Guaranteed maximum price
- Flat fee

Guaranteed maximum price

A **guaranteed maximum price (GMP)** model is a stipulated price that the general contractor will use as their project budget. A variation of this funding model is what is called guaranteed maximum price at risk. A **GMP at risk** means the contracting team is at risk if they exceed the guaranteed price.

There is one more version of the GMP model, and that is a GMP that utilizes the pain share | gain share model (PS\GS). In the **PS\GS** model the contractor(s) will share in the cost overruns (pain) and will share cost savings (gain).

Later in this chapter, I will discuss the pain share | gain share model in greater detail.

Flat fee

The **flat fee** model is where the contractor(s) execute their project's scope for a flat fee.

A flat fee model places more risk on the contractor because the contractor(s) accept the project at a flat fee. If the contractor(s) use a guaranteed maximum price, then there is a set cost that they will not exceed.

Quick note on flat fee and GMP contracts

Depending on the contract language the liabilities of a GMP and flat fee pricing model can be reversed. For example, I have seen projects where the contractor using a flat fee model was not responsible for scope items beyond the price of the flat fee.

On the flip side, I've seen situations where the contractor under a GMP was responsible for scope outside the price of the GMP.

It is important to seek the advice of a contract lawyer before agreeing to any contract.

Schematic design phase

The next step in a construction project is to design the systems for the building. During this phase, the architect and design engineer will work with the project's contractors to design the systems for the project.

Some of the systems that are designed during this phase are:

- Mechanical equipment
- Electrical systems
- Building controls

To be clear, the level of detail in the design at this point and time is purely conceptual. This means that the specific details around building controls, power, and other specific design concerns are not a factor, yet.

You may be wondering, who comes up with these designs?

Typically, these designs are created by a design or a consulting engineer.

At this phase of the project, you have two types of engineers.

The first type is **design engineers** who are focused on designing mechanical, electrical or other specific systems.

The second type is the consulting engineer.

The **consulting engineer**, often called a technology consultant or specialty design engineer, provide design guidance for a specific technology or set of technologies.

A side note for you to consider is that the consulting engineers who are designing the building's systems have different levels of experience. Therefore, you can get different levels of results based on who you select.

In some cases, you can get folks who have a design that they're known for. This is often known as a **boilerplate design**.

I remember several years ago I was working on a rather large high school for a school district. Traditionally with high schools, you'll use air handlers and VAV systems because of the large amount of classrooms and spaces. The air handlers are then supported by a central cooling and heating plant that supplies the chilled and hot water to the units.

So why do I bring this up?

The design engineer for this high school was known for using rooftop units in his designs.

By the time I got involved with the job, the entire high school was covered with 2 to 5-ton rooftop units.

But it got better, so here I am showing up to this cluster of a project, and the design engineer looks over at me and says ok Phil how are you going to design the sequence for the central plant.

Honestly, I was caught off guard. Here I was sitting with the owner, general contractor, mechanical contractor and the design engineer was asking me how I was going to design the central plant?

Didn't this guy know that he was supposed to design the control sequences?

Needless to say, that project was a fun time suck on my life for about six months.

That project ranks up there on the list of the stupidest things I've seen.

Design Phase

Ok, back to the design.

Now that an engineer has been selected the project will move into the Design Phase.

In my experience, there are a couple of ways this process can go.

In the world of building automation companies there exists this delicate balance. You see, if you are a building automation company, then you are always striving to get directly under the general contractor (GC) or, if you can swing it, under the owner.

There are pros and cons to being directly under the owner or GC.

Obviously, the closer you are to the owner the more likely it is that your control system is going to match up with the owner wants. The not so obvious con of this approach is that in some cases there's no filter between the owner and the building automation contractor.

To some of you what I just said amounts to hearsay.

Why in the world would I suggest that having the controls contractor working directly with the owner would be a risk?

The reason why this is a risk is that in some cases the controls contractor doesn't know what they're doing and can provide bad advice. That's why it's important to vet a controls contractor and make sure that they understand design fundamentals, HVAC functionality, low voltage systems, and all the other systems that I described in chapters 1 through 4.

This vetting process is a very important step to discover if the contractor understands the systems that are being installed on the project.

There is nothing worse than a BAS contractor going directly to the owner and advising them on how to deploy building automation systems that they have no experience with.

Selecting a tier 3 contractor

By this point in the project the consulting engineer, architect, general contractor, and all of the tier 2 contractors have been selected. It is now that the tier 2 contractors will begin to select the tier 3 contractors.

The traditional process for selecting tier 3 contractors is for the tier 2 contractors to solicit bids from different subcontractors.

For example, the mechanical contractor may solicit bids from equipment contractors and controls contractors, and the electrical contractor solicits bids from lighting contractors and security system contractors.

Tier 2 contractors will select their contractors based on a variety of factors. In plan and spec projects, which I will cover in just a second, the tier 2 contractor will usually select the contractor with the lowest price.

In design-build or integrated project delivery projects the tier 2 contractors usually select a tier 3 contractor based on the greatest value or most impactful design.

Once all of the Tier 3 contractors have been selected the tier 2 contractors will submit their team's price to the general contractor. Depending on the delivery and funding model the general contractor will select the team with the lowest price or the greatest value.

From this point forward the construction team is formed.

Project delivery models

There are a couple of different models (or methods) that the construction team can use to deliver a project. These delivery models are:

- Plan and specification (PS)
- Design-build (DB)
- Integrated project delivery (IPD)
- Public-private partnership (P3)

Plan and specification delivery model

A plan and specification (PS) project is the most common delivery model in the construction space.

In a PS project, the projects engineers will create a set of plans and specifications.

What are plans and specifications you ask?

Plans, depending on the context, are usually the mechanical, electrical, and plumbing drawings that show where devices will be placed and how systems will be connected.

<u>Think of plans as the blueprints for a project.</u>

Specifications are the specifics of the individual systems that will be installed via the plans. From a controls perspective, specifications detail out the sequence of operations, hardware and software requirements, and the support structure for the BAS system being installed.

Most of the projects I did very early in my career were PS projects. On these projects, I would get a set of mechanical plans and specs from the mechanical contractor. It was my responsibility to look at the plans and specifications to lay out my controls, sensors, and wiring.

I would read then read the sequence of operations that was described within the specifications. I would use this sequence of operations to write the programs for my building automation controllers.

One of the things that bothered me with PS projects was the attitude held by some contractors that the subcontractors were supposed to execute the plans and specifications exactly as they were written.

In a plan and specification project, anything that may not work or may seem wrong is submitted through the Request for Information/ Request for Clarification process (RFI/RFC). The **RFI/RFC** process allows the contractor to ask questions and clarifications around the intent of the design specifications. As part of the RFI/RFC process, the requests will be answered and if necessary changes will be made to the specifications and plans. In some situations, a change order will be issued if the changes are outside of the contractor's scope.

A **change order** is a proposal or set of proposals that seek to rectify the difference in scope between the specified scope and the scope that was agreed upon as part of the RFI/RFC process. There can be negative change orders (where scope is reduced) or positive change orders (where scope is added)

Design assist delivery model

In the **design assist model**, a specialty contractor, like a technology consultant, will assist the contracting team in the design. In this model the consulting engineer or architect will work alongside this specialty contractor.

This team will work to create what is called a design narrative. A design narrative is another term for use cases.

The **design narrative** states what the design team wants to achieve. The design team then looks at how they can achieve the design intent and how the project team can execute the design. Typically, during this process, the project team will go back and forth collaboratively until they come up with the final design narrative.

That design narrative is then submitted for approval.

From this final narrative, a guaranteed maximum price is created to help the project team figure out a budget for the project. This GMP is submitted, and then from this GMP you'll move into the construction project.

One concept that a BAS contractor needs to be aware of is the concept of the "Engineer of Record."

As a BAS contractor, when you are assisting in the design you should not be performing any engineering duties. It is important for a BAS contractor to avoid assuming the role of the Engineer of Record. If the BAS contractor is deemed to be responsible for the design, then they may be held liable for any design failures.

Ultimately this is determined by how the contract is written.

Integrated project delivery

Integrated project delivery (IPD) is a newer project delivery model that is starting to catch on.

In an **IPD** project the general contractor, subcontractors, and engineers form a team. This team will work with the owner very early in the construction process to come up with the design.

Some IPD models support what is called the pain share/gain share model that I covered earlier in the GMP Pricing model.

Once selected, the IPD team will work together towards a budget.

If the pain share | gain share financial model is used, and the team goes over this budget, then the team will share the pain associated with any overrun. Cost overruns can be weighted towards the contractor(s) who caused the overrun, or they can be weighted towards the entire team.

This is determined by the individual contracts that were signed by the contractors. If the IPD team executes the project under budget they can share in the benefit, or gain, of that budget surplus.

Public-private partnership (P3)

The final project delivery method is the public-private partnership or P3 model. This model used to be seen mainly in Canada I am starting to see this model being used in the United States. There seems to be a reluctance to use this model because the model is essentially a lease model.

In a P3 project a team, often called a **consortium**, will create a design model that covers both the construction and operation of a building. Typically, this operational period is 30 to 50 years long. The consortium will also come up with the financing mechanism for this project.

Once a consortium is formed it will submit its proposals, which contain the team's designs and costs, to a selection committee. This selection committee will pick the consortium that will execute the project.

As I stated earlier, the P3 model is funded by a private organization. This private organization funds the construction of the building. Depending on how the financing is set up the owner will lease the building from the private organization or the team.

An example of this model that you may be familiar with is toll roads.

What you may not know, is that private companies build many of the toll roads in the US. The private companies can lease the ability to use the toll road to the state for a period of time. In most cases, it is the responsibility of that private organization to operate the toll road.

P3 construction projects are similar to toll roads, in that the team will get funded by a private lender. This private lender is investing in the building. Because this is a long-term investment the selected consortium will be required to design the building with a detailed lifecycle model.

A **lifecycle model** determines the operational cost of a building over its projected lifecycle.

Essentially a lifecycle model states that a building will have operational costs that are within a range. Some of the costs that are accounted for are:

- Utility costs
- Maintenance costs
- Staffing costs

In some cases, the costs of technology refreshes and equipment replacements are included as well.

Based on these costs the consortium will have a packaged cost that consists of the construction budget and operational budget.

These budgets are subject to the same pain share/gain share model as the IPD project with one major difference. The PS/GS model covers the operational costs as well. I am getting a bit deep into the P3 model, but I feel it is important for you to understand how the costs breakout in case you are ever involved in a P3 project.

My final point, on the P3 model is that the contractor(s) who are involved in the construction of the building are usually responsible for operating the building over that its operational lifecycle. As you can imagine this is a huge amount of predictable revenue for the executing contractor(s).

Design phases

Once the delivery models, funding models, and contractors have been selected the project will move into the design phase. There are three primary design phases in a construction project:

- Schematic design
- Design
- Construction design

Schematic design

The purpose of the schematic design is to produce the schematic design drawings (often called SD). Often the SD's are the first drawings created that involve the contracting team.

During the schematic design phase, a very high-level design is created. This high-level design can be within 20% to 30% accuracy of the final budget.

This purpose of this phase is to flush out the design narrative. For example, during this phase, the design engineer could determine that the project needs an air handler, chiller, and terminal units. This equipment would be added to the SD.

However, the tonnage and flow rates for these pieces of equipment will not be determined yet. The purpose of the design at this point is to place stakes in the ground that define, very generically, what systems will be in the building.

Design

The design phase seeks to produce design drawings, often called (DD). The DD are the drawings that begin to solidify the design.

Design drawings are created to help the contracting tiers define the functional capabilities of the equipment that will be used on the project.

In the SD phase, the type of mechanical equipment was defined. In the DD phase the specific controls, equipment, and design parameters are selected. For example, at this point in the project, the details of individual products and their functional characteristics are defined.

Construction design

The construction design phase is where the construction drawings, called (CD), are created. CD's define how the building will be built and how the individual systems will be installed.

The CD will answer questions like:

- How will the building automation system be wired?
- How will the HVAC equipment be installed?

The specific installation details are detailed out in this phase. CD's will be the drawings that are used during the construction project.

The project process

A construction project will begin with a meeting called the project kickoff meeting. This meeting should not be confused with the initial scoping meeting that I covered at the beginning of this chapter.

The **project kickoff meeting** will usually happen when the schematic drawings are completed. By this point, the general contractor has issued contracts to the Tier 2 contractors. The project kickoff meeting serves to solidify the meeting cadence, safety regulations, and the project schedule.

It is now that the contracting team will begin to execute the project.

What follows next is an abbreviated version of how a typical construction project is executed.

A new construction project will usually start with the excavating of the building envelope. It is within this area that the foundation will be poured.

From this foundation, the support structure or core of the building is built.

It is at this time that the projects contractors will bring in any utilities, like the main electrical feeds, plumbing, sewage, etc. During this part of the project the building automation system contractor is usually just starting to finalize their design drawings unless they been involved very early in the design.

If this is a design, assist project the specialty contractors will be working to help create design narratives.

Once the foundations are laid, and the core of the building is underway, the other contractors will begin their work. Electricians will start to run wiring, plumbing contractors will begin to run piping, and mechanical contractors will begin to install duct work.

The building automation contractor will start to staff, or man, the project.

The first thing the building automation contractor will do is walk the site and confirm the locations of sensors and panels with the contracting team. This is done before installation to ensure where to run the controls wiring, install the controls panels, and mount the controls sensors and devices.

To be clear, though, at this point, the BAS contractor is not mounting thermostats or installing things. Rather the BAS contractor, usually through their electrician is roughing in their devices.

Roughing in devices is where the BAS contractor or electrician run wires to their sensor locations to "rough in" the sensor locations. This ensures the wires are present and available once drywall or other obstructions are installed.

Once the core and the shell of the building are done, the BAS contractor will start to connect their sensors and program their devices.

At this point in the construction process, the building usually has power, and the mechanical systems are installed. The BAS contractor will work alongside the mechanical contractor, air balancer, and commissioning agent to test the BAS controls and mechanical equipment. This process is called **startup,** and it is during this process that the equipment is checked for basic operation.

Next, depending on the project, the BAS contractor, and the commissioning agent will work through functional testing. Depending on the scope of the project functional testing may or may not need to be required.

By the way, it is at this point that things tend to get very chaotic.

On poorly planned projects the project schedule can call for the BAS contractor to do startup tests, functional tests, and air balancing at the same time, which as you can imagine is a little bit crazy.

I'm going to dive into each one of these topics. After I do that, I will discuss why doing each of these steps at the same time is a bad idea.

Startup Testing

Startup testing is where the BAS contractor and other contractors are testing that the equipment will start up and that the commands from the BAS work.

A classic example of this is VAV box testing. This is where the BAS contractor is working alongside the mechanical contractor, tests that all of the VAV damper actuators are stroking.

Balancing

Once the systems are operational the balancing agent will begin his/her work. In chapter 1 I discussed how there is a finite amount of air or water coming out of your HVAC systems. Because of this, you need to ensure the air/water is being delivered to the right places.

Balancing is the process of adjusting the airflow or water flow at various points in the building. This is done to ensure that the mechanical systems are delivering the amount of air or water they were designed to deliver. The person who does this is known as a **balancer**.

The balancer will use balancing dampers and balancing valves to manually adjust the system. The balancer will also work with the BAS contractor to ensure that values in the BAS match with the values he/she sensed at the various sensors (airflow, temperature, pressure, etc.).

This is called sensor calibration.

Functional Testing

Now that the startup is done and the system is balanced it is time to perform functional testing. As I said earlier, not all projects include functional testing.

Functional testing is when a commissioning agent or the BAS contractor operates the BAS system to drive the equipment that the BAS controls through its sequence of operations.

Not all projects have a commissioning agent and the projects that do have one may not have the commissioning agent perform "full commissioning."

In my experience, there are three levels of commissioning:

- Design review commissioning
- Sample testing
- Full functional testing

Design review commissioning

In **design review commissioning**, the commissioning agent will review the design documents and advise the construction team of any changes or areas of concern that the commissioning agent has.

Sample testing

Sample testing is where the commissioning agent will take a random sample of a certain percentage of systems. The commissioning agent will validate that this sample pool meets the functional requirements of the sequence of operations.

Full functional testing

In **full functional testing** the commissioning agent will go through and test all the systems to that make sure are operating according to their design.

Why Doing Startup, Balancing, and Functional Testing at the same time is a bad idea

Earlier I mentioned that doing the startup, balancing, and functional testing at the same time was a bad idea.

This is because of how each of these different steps flow into one another. If you think about the logical flow of each of these steps, the first step you need to take is to make sure that the systems will start.

After all, you can't balance a system or check its functionality if it won't start. I have been on so many projects where the startup of the equipment was not complete. Not only does this waste folks time, but it also ends up pissing people off, which can come back to bite you in the butt later in the project.

After startup has been performed on the equipment and controls the system needs to be balanced. There is a very important reason why you'd want to focus on balancing the system before testing its functionality.

The reason behind balancing a system before performing a functional test on it has to do with downstream dependencies.

For example, if you have a central cooling plant that is providing chilled water to all of your other systems you cannot verify that the system is controlling to its set point if parts of the system are not getting adequate chilled water.

Hopefully, now you understand why the startup, balancing, and commissioning should not be done at the same time.

As-builts and close-out documentation

As-builts and close-out documentation is an area I have been particularly interested in due to how many projects I've seen fail in this area.

As-built and close-out documentation are critical documents for facility operations. However, on a lot of projects the construction team is over-budget and in a rush to get off the project. This results in as-builts that are simply copies of the construction drawings.

So what are as-builts?

As-builts are documents that show the condition of the systems as they were built, hence the name as-builts. As-builts, also known as red-lines, describe out how the systems were wired, how the sequence of operations was programmed, and how the different systems were installed.

Close-out documentation, on the other hand, can be seen as final submittal documentation. Early in the construction process, the contractors submitted their submittal packages. The close out documentation is essentially the same process except that it includes the documentation (installation manuals, catalog pages, etc.) for the systems that were installed.

Final training

Training is usually the final hurdle in the construction project. Since there is no standard that defines training the quality of training is all over the map.

That is why when you are either scheduling training, specifying training, or attending training you should require a solid training agenda.

I go into great detail on how to create a training program in my _Audio Course How to Evaluate a Building Automation System_.

Warranty and operations transition

At this point of the process, you can see the light at the end of the tunnel.

The final step of a construction project is the warranty phase. The **warranty phase** is typically a year-long period during which any "warranty" issues related to the new systems are supported.

As you may have noticed, I put the word warranty in air quotes. The reason I did this, is that some folks, I'm sure you're not one of them, will milk the warranty process for free work and service.

As you would expect, the warranty process can be an area of contention between the contracting tier and the owner if the warranty scope was not well defined.

After the warranty period has concluded the contracting team will hand the project over to the owner who will then operate the building.

Chapter 8 Quick Summary

There were several key takeaways from this chapter, and I wanted to make sure I highlighted them for you.

The first key takeaway is understanding the funding and delivery models that are being used on the project. Each funding and delivery model affects your role in a project differently.

For example, I have seen owners who want to have the ability to customize their systems but because they selected a plan and specification delivery model they find that very difficult.

The next thing I wanted you to get from this chapter was the importance of planning.

Without proper planning, you can experience serious execution issues. These issues tend to show up late in the project. An example of this would be when a BAS contractor is not aware that commissioning is part of the project scope. Because of this, the BAS contractor does not allocate proper costs to his/her bid.

The final thing I want you to take from this chapter is the importance of proper project close-out. It is extremely important for you to ensure that the as-builts are created in a manner that supports the operation of the building post warranty. In addition to this, it is very important to have an effective training agenda agreed upon prior to executing training.

Sometimes you don't need to build a new building. Sometimes, your BAS is just a dinosaur that needs to be put out to the pasture…

Well, in the next chapter, I am going to discuss those "sometimes" and how to handle them.

CHAPTER 9

WHEN SHOULD YOU UPGRADE YOUR BAS?

What's in this Chapter

Whan should a building automation system be upgraded?

A lot of folks I meet have this exact question. The challenge with upgrading a BAS is figuring out exactly where the tipping point is.

What do I mean by the tipping point?

The tipping point is when an existing BAS becomes significantly more expensive to maintain than the cost of upgrading. This brings me to my next question.

How do you know when you have reached a point where your system needs to be upgraded?

In this chapter, I am going to discuss both major upgrades and the often ignored minor upgrades. I am also going to talk through firmware (that is a fancy way of saying controller software) upgrades.

So here is what you are going to learn about in this chapter:

- Making the decision
  - o How do you know when your BAS is too old?
  - o What parts of your BAS should you upgrade?
  - o Monetizing the risk of not upgrading
- Planning the upgrade
  - o How do you plan for an upgrade?
  - o What can you expect on the day of the upgrade?
- Advanced topics
  - o How do you handle multi-vendor upgrades?
  - o How do you maintain integrations when you upgrade?

Ok, are you ready to answer the question that has been plaguing both owners and contractors for decades?

It's an ambitious goal, but by the end of this chapter you will understand:

- How to make the upgrade decision
- How to plan upgrades
- How to handle complex upgrades

Alright, let's dive in!

Making the decision

Let's face it upgrades are scary. Whether you are the customer or the contractor, there's a lot of risk in upgrades. Often, even the customer doesn't know what is in their building.

So there you are, trying to figure out how to upgrade a system, while also trying to convince the customer (sometimes this is internal customers) that this upgrade is necessary.

And that is why this section is first.

At the end of the day, your main goal should be to understand why the upgrade is happening and the benefit that you or your customer will see because of this upgrade.

I argue if there is no benefit to an upgrade why upgrade?

Don't worry. I'm going to cover that question as well.

So let's start off with the age-old question, how do you know when your BAS is too old?

How do you know when your BAS is too old?

I've built my whole career around upgrading and integrating building automation systems. It never fails to amaze me some of the ways my customers were able to keep their systems running long beyond their prime.

I still remember the first time I got shocked by 120V, trying to retrofit a pneumatic system that was on its last legs.

No matter what I tried, I just couldn't get the fan on this unit to turn off. So there I am ripping out wires and cutting out pneumatic tubing, and I see this alligator clip holding two wires together. Because the wires were so small, I thought that they were communication trunks.

So what I do?

Well, I just reached out and grabbed that alligator clip and pulled it off the wire. I then proceeded to have my arm thrown back into sheet metal enclosure that I was standing inside of.

Man, that sucked.

And that is how I learned that this jury-rigged building automation system had outlived its useful life.

When you have to start alligator clipping wires together or fusing pneumatic tubing together just to get things to work, it might be time to consider replacing that system.

But I know you all, you're not get a let me out of answering this question that easily.

No, no, you all want a checklist on how to determine when your BAS to is too old.

Well, ask, and you shall receive. Below are the four questions I ask to determine if the BAS is too old:

1. Can I no longer buy parts for the building automation system?
2. Is the building automation system manufacturer out of business?
3. Can I no longer find anyone who knows how to support this building automation system?
4. Am I spending more to maintain my existing system than it would cost to upgrade the system?

Whether you are a contractor or an owner you should stop and ask yourself these questions.

You see, upgrading a building automation system should not be an emotional decision. In my experience, the decision to upgrade a BAS is made based on the answers to these four questions. The answers to these questions will help you determine if a building automation system is no longer viable.

What parts of your BAS should you upgrade?

At this point, you probably have one of the following questions.

The questions go something like this.

What parts of the building automation system should I upgrade?

I mean, surely you don't expect me to upgrade the entire thing?

Or

How do I know if the entire building automation system needs to be upgraded?

It depends on how you answered the questions from the previous section.

For example, if you have a BACnet MS/TP communications trunk and the field controllers can no longer be bought. Then it makes sense to upgrade your field controllers with new field controllers and leave your supervisory controllers and servers in place.

The reason to upgrade the field controllers is because BACnet MS/TP is a common protocol and there are tons of controllers that support it.

However, if the BAS servers and supervisory controllers can no longer be bought or they use a proprietary protocol, it may not be possible to replace the field controllers.

Therefore, this problem should be approached on a case-by-case basis. There are three parts of the building automation system that an upgrade project typically focuses on. You may recall these from Chapter 2:

- Building automation server
- Supervisory device
- Field controller

Here are four questions that you can ask to determine which parts of the BAS you should upgrade.

As you ask these questions, it is important to start with the building automation server and work your way down to the field controllers. As you move through each layer continue to ask yourself or your customer the following four questions:

1. Does this software support open protocols?
2. Are these open protocols present in products on the market?
3. Can I still find people who will service this system?
4. Can I still purchase the system (this is mainly focused on the supervisory devices and field controllers)?

As I said, you will continue to ask these four questions as you move down from the supervisory device to the field controller.

Whenever you discover that the answer to one of these questions is no, you need to make a decision about whether or not you want to take the risk of continuing to use that system.

Monetizing the risk of not upgrading

How do you determine the risk of not upgrading?

In chapter 12 I will describe how to determine a return on investment (ROI) for service activities. I've included the formula from chapter 12 below.

$$ROI = \frac{(\text{Gain from Investment} - \text{Cost of Investment})}{\text{Cost of Investment}}$$

You can follow the ROI equation and process from chapter 12 to determine the return on investment for upgrading, with some minor tweaks:

- First, you need to determine the cost of downtime. You do this by looking at what areas and functions of you or your customer's building and business, could be impacted by system failures
- Then, you will determine the cost of upgrading the system(s)
- Next, you will use the ROI formula to determine if the ROI justifies the cost of the upgrade
- If the cost of upgrading the systems meets your criteria, then you can proceed with the upgrade.

Planning the upgrade

At this point, you should have answers to the economic, technical, and operational questions related to this upgrade. You should also have evaluated the ROI for this project and decided that an upgrade is needed.

Okay, now what?

I've met a lot of folks who get to this point and freeze up because doing an upgrade can be pretty darn scary. I remember the first upgrade I ever did, and it didn't go well.

I went to a Native American hospital about three hours north of Seattle. I was supposed to replace a Tridium LON supervisory device with a new supervisory device and about a dozen new programmable field controllers.

Well right off the bat I set myself up for failure by not grabbing a copy of the database. My thinking at the time was that since I was only adding some new controllers, I wouldn't need a backup of the database. (Hopefully, you can read the sarcasm in my words) …

After going and taking the Tridium device off the wall and putting my supervisory device on the wall, I proceeded to wipe out the LON database completely. This led to the extremely fun process of spending the next three days at this hospital rebuilding the database from scratch.

To this day I'm still amazed that I didn't get fired, but in my defense, I've heard much worse stories than mine. The point is, I didn't have a plan, and that was pretty dumb.

But, you will have a plan, because I'm about to give you a step-by-step process for creating one.

How do you plan for an upgrade?

Yes, I'm talking to you, you've got to have a plan!

What follows, is a step-by-step plan that will help you to avoid the mistakes and pain that I experienced in my career.

You ready?

There are five steps, and if you follow them, they will make your upgrade projects much less painful.

Step 1: Determine the scope of your upgrade

What are you going upgrade?

Are you going to upgrade all the controllers in a building?

Just some?

Will you need to have a new communications trunk?

When you approach an upgrade, you will need to consider, the physical, logical, and operational impacts to a system. You also need to consider if your upgrade has the potential of impacting a part of that system.

I like to check the following areas:

- Will this impact the database?
- Will I only be replacing a couple of controllers on a communications trunk?
- Will I have communication speeds or protocols that could conflict?
- Will this new system need to work with the old system?
- Will I need to manage multiple user databases?
- Are there sequences of operations that are dependent on points from the controllers I'm upgrading?

This is not an exhaustive list, and it really will depend on what you are upgrading, but the list above should give you a head start.

The important thing is to check the physical, logical, and operational parts of the system(s) you are upgrading.

Step 2: Pick the order of systems you will upgrade

Now I realize you won't always start with a specific device. In some cases, you'll start with the supervisory device and other situations you may start with a field controller. The point here is not what device you start with. Rather it is the order in which you proceed to perform your upgrade.

As I see it, there are three schools of thought in when it comes to planning out an upgrade.

Method #1: Start with the field controller

In **method one** you start with the field controller and then gradually work your way up to the supervisory device. The benefits of this approach are that you only have to build the new database once. The cons to this approach are that you will have to go and run your systems in hand (meaning in manual override) until all of the controllers are upgraded.

Method #2: Start with the server or supervisory device

Method 2 starts with the server or supervisory device and flows down to the controller. The thought process is that you will go and stand up the new supervisory device with all of your field controllers in hand. Next, you'll start replacing the old field controllers, and as you replace the old field controllers, you will map your new field controllers into the new supervisory device.

Method #3: The hybrid approach

Method #3 is the hybrid approach. In this approach, you will take a parallel path where you are upgrading the supervisory device and the controllers at the same time. This approach has the benefit of being faster and minimizing the time systems are in hand. However, this approach is more complex and requires significant planning and expertise.

Step 3: Plan your rollback strategy

Okay, let's say your whole plan goes to crap what you do?

If all the sudden the controllers don't work or you somehow missed a linkage point between controllers, how can you get everything back to operational as possible?

In the IT world, this is called a business continuity or disaster recovery plan. Quite simply, the continuity strategy will detail out each change and the rollback method for that change.

For example, changing out a supervisory controller would be an individual change.

The rollback method for that change would be to reinstall the old supervisory controller.

In this step, you need to lay out each change and the appropriate rollback method for each change.

Step 4: Plan out your project schedule

Ok, now I'm going back to project management basics.

Quite simply you need to identify each task and the prerequisites for each task. Once this is done you need to assign personnel and timing to the tasks and work your way backward from the targeted completion date to the start date.

I'm not going to teach you project management, and resource mapping in this book. Just realize that you need to plan out your project just like any other project.

Step 5: Define your commissioning or startup process

Alright, the final step.

In this step, you will detail out how you will validate that the new system is functional. To do this, you will need to create a series of functional tests where you will go and check that the sequences and systems are operating as expected.

There are three ways to do this:

- You can spot check a subset of the devices that were upgraded
- You can hire a commissioning agent, which I covered in chapter 8, to validate that the new systems operate as designed.
- You can self-execute a full functional check on the systems.

I do not recommend method #3, as folks tend to get busy and never complete the functional testing.

How to execute an upgrade project

Okay, so now that you feel comfortable with planning an upgrade project it's time to talk through how to execute an upgrade project.

Now, I want to be clear that no execution plan will ever be 100% accurate.

Therefore, I do encourage you, to customize the 12 steps I am about to walk you through to fit your project.

With that being said, what I am covering in this section will form the bedrock of any upgrade project execution plan.

Based on my experience, there are 12 steps that you will want to take in order to execute your upgrade project effectively.

These 12 steps are:

Step 1: Decide on what day you will complete the project

I know this step seems obvious. Trust me. You'd be surprised how many upgrade projects are started with no defined completion date.

Often the owner, who may have never been involved in an upgrade project, doesn't feel comfortable asking for a date because they don't understand what is involved.

Owners, it's ok if you don't know!

You don't have to understand what is involved in asking for a date.

And contractors… pick a date!

Yes, I know things change and stuff pops up. As you move through these steps, you might find out things that make you adjust your date.

DON'T WORRY

There's an easy way to handle that, simply state something in your contract that says *"your current competition date is based on the information you have gathered so far and that the date is subject to change based on site conditions."*

Once again, I'm not a contract attorney so before using any specific contractual language make sure you get a legal expert to evaluate your contracts.

Step 2: Verify the job site systems and applications

If you have been following along up to this point you should have a list of the systems, you will be upgrading. Now you need to verify that the list of systems and applications that you will be upgrading is accurate.

This is a very simple step, but it can be time-consuming, so budget accordingly!

What you want to do, is to go to each system and verify the following:

- The location of the system
- The location of the systems manual override (if one exists)
- Any documentation related to the system
- Any code related to the system (if you are doing a software upgrade)
- The electrical and mechanical systems that supply or are supplied by the system

As you may imagine it can take a lot of effort to go and identify this information.

You may be tempted to skip this step. After all, you are going to be touching the systems anyway so why would you need to get all this information?

I'm here to caution you to avoid the temptation to skip this step.

I can speak from personal experience; any time I have skipped the information gathering portion of an upgrade I have ended up regretting it.

This regret can manifest itself in a variety of forms. One form of regret was labor that was unaccounted for due to the difficulty to access the system.

Another consequence of not properly collecting and verifying information is the additional costs related to not being able to upgrade a system due to missing software or configuration tools.

To make a long story short, make sure you verify the information I listed above before upgrading a system.

Step 3: Ensure you have the access you need

Access is critical for most upgrade projects.

But access to what?

I mean, we've already gone and identified the location, documentation, and code for the systems.

What else could you possibly need?

Well here is one huge area that tends to be missed. This area is the area of credentials.

What are credentials you ask?

Credentials, also known as usernames and passwords, allow you to access the applications or software for the systems that you're upgrading. Imagine that you are doing an upgrade to a system that contains programming to control other systems.

Would you just go and replace this system without understanding the code inside it?

Imagine the potential damage you could do if you just took wild guesses as to how the building automation system is performing its discharge air reset, optimal start, and zone averaging sequences?

I think you would agree that the potential damage you could cause to your systems is quite high if you don't understand what they are currently doing.

But you can't understand what your systems are doing if you can't access the systems.

You don't want to find out, two weeks into a project, that the systems manufacturer will not provide you access to the system and you have to guess as to how it functions.

Don't get stuck in this situation, make sure that you have the access you need before executing your upgrade project.

Step 4: Identify the systems that will be affected

But Phil, I've already gone and identified the systems that will be affected.

Did you?

Are you sure?

Often time's folks will only look at the systems they're upgrading, and they won't look at the systems connected to the systems they are upgrading.

For example, maybe you're upgrading the controls in a central plant. That's great because there are some awesome control sequences for central plants that have come out in the past couple years.

However, if you don't think about the systems that tie into and use the central plant you could end up with significant impacts on those systems.

Maybe you have a secondary chilled water loop that flows into another part of the campus. And this loop has some code that averages the valve position of your air handlers and resets a set point based on the average valve position.

Now imagine that you didn't know that this code existed. Because of this, the person who goes and writes the new code for the central plant creates a different way of controlling the secondary cooling loop.

Could this introduce problems?

Possibly.

Do you want to find out after the projects done?

Probably not.

So what can you do to avoid this scenario?

Well, I recommend that you go and identify the systems that will be affected or tie into the system you're upgrading.

Once you've identified these systems, you need to figure out what aspect of the system could be affected by the system you're upgrading. Once this is done you need to put a plan in place to ensure that the system(s) is not being affected by the upgrade.

Step 5: Decide on your upgrade strategy

In step two of the upgrade planning process, I described the different methods you can take when you perform an upgrade.

It's at this point that you need to select the method you will use. You will use that method to determine the people, tasks, and timing you need to put in your execution plan.

Step 6: Determine the people or groups that will be involved

This step is pretty basic.

By this point you should have a list with the following information:

- The systems you will be upgrading
- The systems you need access to
- The systems that will be affected by the upgrade

Based on this information, you will need to determine the people and or functional groups that need to be involved in the upgrade.

I recommend that you break this list of people into three groups. These groups are:

- Advise
- Approve
- Contribute

Advise

People who are in the advise group, simply need to be advised on any potential impact to their workspace or their systems.

An example of this would be notifying the head nurse when you're about to do an upgrade in a patient room.

Approve

People in the approve group will approve the actions you're taking and the allocation of resources to your upgrade project.

Contribute

People who are in the contribute group will be utilized to assist you in performing the upgrade. An example of this would be a facilities team member who needs to put a specific piece of equipment in hand for the upgrade to be executed.

Step 7: Write out your upgrade plan

At this point, you should have everything you need to write out your upgrade plan.

But if you've never written out an upgrade plan how do you do that?

In the bullets below I have provided you the eight areas that you will need to address to form your upgrade plan.

- Upgrade scope:
- Point of contact:
- System(s) to be upgraded:
- System(s) affected:
- Upgrade steps:
- Upgrade period:
- Completion date:
- Rollback plan:

You need to fill out each one these areas with the information that you gathered in the previous steps. You will need to complete these areas for each individual system.

Once you have these areas completed, you simply print out this form and provide it to the customer and post it in the area where you're working.

Step 8: Put the systems in hand

In step seven, you detailed out your upgrade plan.

One of the areas you filled out in this plan was the upgrade steps. One of the steps you should have, depending on the type of upgrade you're doing, is to put your system(s) in hand.

I highlighted this step because I have been on upgrade projects where the controls personnel did not put the system in hand. Because of this, the central cooling plant began to lose control, and this resulted in a loss of cooling to one of the buildings on a college campus.

Upgrades are complicated enough. You don't want to add further complications by leaving your systems in automatic control.

Step 9: Execute your plan

This is the moment you've been waiting for.

All of your planning, preparation, and research has led up to this moment. Now all that is left is for you to execute the plan you created in step seven.

There are a few important things to remember when executing an upgrade plan. These things are:

- Keep a copy of your upgrade plan near where the upgrade is taking place
- Practice your rollback scenario (if you have one) before executing the upgrade
- If appropriate put your systems in hand before upgrading them
- Notify the appropriate personnel when you begin the upgrade when you finish the upgrade, and if something goes wrong during the upgrade

Step 10: Document your changes

After your upgrade is complete, you will have to provide updated documentation.

This step shows how the upgraded system works and details out any changes that were made. This will assist with servicing the system in the future.

Specifically, you're looking for:

- Sequence of operations
- Schematics
- Wiring and flow diagrams
- Bill of materials
- Cut sheets on any new material

Step 11: Verify proper operation

Once the upgrade is complete, you have two more steps before you're done.

Depending on the kind of upgrade you performed, you may have changed or replicated the previous sequence of operations. It is important that you validate that the new or previous sequence of operations function as designed.

This is a pretty straightforward process where you do the following:

- In the case of a controller upgrade, verify all new and existing inputs and outputs still work
- If you were performing a supervisory device upgrade, ensure that your communication trunks still work and that you can still communicate with your controllers. Also verify that any, graphics, trends, alarms, and custom logic still functions as designed
- If you are performing a server upgrade, then verify that you can access the server, run your applications, and perform any server specific tasks that are required

Step 12: Check the opposite season control

This task does not have to be done on every upgrade project.

However, I encourage you to make sure that the systems you're upgrading do not have a different sequence of operations based on the time of year or season of the year.

An example of this would be a two-pipe system that has a summer-winter changeover, to switch between cold and hot water.

If a system does have a different sequence based on the time year, then you should plan to perform an additional functional test to verify that the upgraded systems function properly.

Advanced Topics

What I have covered so far will cover 95% of the upgrade projects you will be involved in.

However, there are two of aspects of upgrades that I want to cover in further detail.

Now I caution you. These are advanced topics, and it's perfectly fine if you simply move on to the next chapter.

In this section I'm going to cover:

- How to handle multivendor upgrades
- How to maintain integrations when upgrades are performed

How do you handle multi-vendor upgrades?

The reality is some upgrades will involve multiple different building automation vendors.

I experienced this back in Dallas, Texas, when I was working on upgrading the control of a central plant that was tied into another building automation system. What made this even more difficult was that building automation system had been purchased by another company and was no longer supported.

In order to support this new central plant control sequence, I had to make changes in the existing third-party system.

In this section, I will tell you how I addressed that and give you a step-by-step process for handling multivendor upgrades.

Much of what we cover in this section will be similar to our previous steps around planning and executing upgrade project. I am going to call out the differences rather than running through the entire process in full again.

There are three main areas you need to focus on when doing multi-vendor upgrades. These areas are:

- Identify the third-party systems that will need to be upgraded
- Ensure that you have access to the third-party systems
- Hire or contract someone who understands the third-party system

Identify the third-party systems that will need to be upgraded

You may recall that in step two and four of the upgrade project execution section I detailed out how to identify the systems that will be upgraded and affected by the upgrade. When dealing with a third-party system you simply need to perform the same tasks on the third-party system.

The only issue with this is you may not have the ability to access the third-party system or the expertise to work on the third-party system.

Ensure that you have access to the third-party systems

There are two ways to ensure that you have access to the third-party systems you're going to be upgrading.

The first way is to hire or contract someone from a third-party system to ensure that you have access to and the software for the system. This will allow you to perform your upgrades.

The second way is to work with the customer or your internal team (if you are the customer) to utilize personnel who are familiar with the system to determine whether you have access to the system.

Hire or contract someone who understands the third-party system

Okay, I've said several times now that you need to hire or contract someone who understands the third-party system but how exactly do you do this?

This is probably one of the most difficult areas when performing upgrades that involve multiple different vendors. The reason behind this is that upgrades are inherently risky, and the assignment of this risk is often disputed, especially between competing vendors.

So how can you go and hire someone, especially if that someone is your competition?

The best method I have found is to work with the owner. Obviously, if you are the owner, then just follow the steps below.

Step 1: Identify the people with the capabilities you need

This is quite simple.

You simply need to, identify the tasks that need to be done on the third-party systems and then based on that information you will need to contract the appropriate people.

Step 2: Define the scope that these people are going to perform

Before contracting these people, you will want to design a clear scope around what they will be doing. In the scope you will want to detail out:

- What systems they will be working on
- What tasks that will be performing
- When do these tasks need to be complete
- Who is responsible for the tasks
- The communication and escalation of issues related to the tasks
- The result or finished product that is expected after the tasks are done

Step 3: Contract the people to perform the work

How you contract people will depend on your organization.

The important thing is to have a separate contract for any work that involves a third-party system. The reason behind this is that the terms and conditions of the scope of work can be quite different from what you or your contractor has been hired to do.

I am not a legal expert, so I encourage you to seek out legal or professional advice related to the formation of contracts and subcontracts.

How do you maintain integrations when you upgrade?

While I will cover this topic in much greater detail in chapter 15, I still wanted to take a brief moment to mention a couple of key things.

First, you need to make sure, that you follow step 10 of the upgrade execution plan process. If you recall, step ten is where you document the changes to the system or systems when you're complete with the upgrade process.

As you may recall, the information you need to gather after the upgrade process is complete is:

- Sequence of operations
- Schematics
- Wiring and flow diagrams
- Bill of materials
- Cut sheets on any new material

But you need to gather couple more things to ensure that you have the proper information to maintain your integrations.

First off, before executing the project, you need to gather the physical and logical diagrams that detail out the existing integrations.

Next, you need to determine how, if at all, the integration(s) will change.

Finally, you need to test the integration(s), to verify that the integration(s) still function as designed after the upgrade.

As I mentioned, I will discuss integration in much greater detail in Chapter 15.

Chapter 9 Quick Summary

Upgrades, upgrades, upgrades...

This is one of the toughest topics for folks, and I understand why. You have so many competing emotions when you are going down the upgrade path.

Is the BAS contractor trying to sell me something I don't need?

How can I convince the customer that not upgrading is a very bad choice?

How can I upgrade this site that the customer has let languish for 30 years?

All of these questions strike at the heart of what makes upgrades challenging...

In this chapter, my goal was to make this scary area a little "less scary" for you.

I walked you through how to make a decision when it comes to upgrading. I showed you how to answer the questions around when is your BAS to old and how to identify what parts of a BAS you should upgrade.

I also taught you how to monetize your upgrades as well.

I then pivoted to the operational side of upgrades where I walked you through my step-by-step process for performing upgrades. I even showed you what to expect on the day of the upgrade.

Finally, I took you through some advanced topics like multi-vendor upgrades and upgrading integrated systems.

All-in-all I gave you a pretty solid framework to build off of for your upgrade projects.

And that completes section 2. Section 2 focused on the art of procuring, building, and upgrading BAS. In section 3 I'm going to take you through how to support the BAS you have.

I'm going to give you several step-by-step processes to guide you through managing your BAS!

SECTION III – AFTER THE TRUCKS LEAVE, SUPPORTING YOUR BAS

Section Overview

Now that you know what a BAS is and you've got one installed its time to discuss how to support and manage a BAS.

In this section, I will be walking you through the process of managing a building automation system. I will show you the primary tasks that a BAS is used for and I will give you processes and checklists to help you better manage you or your customers building automation system.

CHAPTER 10

MANAGING BUILDING AUTOMATION SYSTEMS

What's in this chapter?

Managing building automation systems, where do you begin?

Managing a building automation system can seem overwhelming. After much thought and research, I came up with three topics that will make this task easier to do.

In this chapter you will learn:

- The four main ways facility teams use BAS systems
- Six common tasks that are performed using a BAS
- How to create effective standards for your BAS

With this understanding, you will be able to design, implement and operate BAS systems more effectively.

There's a ton to this chapter and the best way to unpack it is to dive right in.

Let's begin!

The four main ways BAS are used by facility teams

Before I wrote this book, I went and met with several facility managers. When I met with these folks, I asked them, "What are the key functions they expect their BAS to perform?"

I narrowed down their responses into four key tasks that a BAS should be able to perform. These tasks are:

- Control building systems
- Troubleshoot performance issues

- Maintain current building system performance
- Optimize overall building system performance

In this section, I describe what each function is, and the common tasks in each of the four functional areas. Where it makes sense, I will line out process maps that you can use to work through each of the functional areas.

Control

At the end of the day, the typical building occupant just wants a comfortable, safe building. To provide that, you need to have control of your building systems. That is why the first function I will cover is how to control your building systems.

Building systems can get rather complex, but at the core of any control scenario, you have some common functions that are required. At a high level, the control of building systems is accomplished in two ways.

These ways are automatic and manual control.

Manual Control

Anytime you need to manually switch over your power feeds or turn on your lights via a light switch you have a manual system. The level to which the system is manual operated will vary. Having a manual system is not necessarily a bad thing.

For example, if you run a small commercial office building you probably don't need to automate the individual control of each room's light via a smart lighting system.

Manual control typically exists at three levels. These levels are:

- System level
- Controller level
- Device level

I am going to illustrate this using an electrical system.

At the system level, manual control can be accomplished through a manual transfer switch at the main circuit breaker.

Manual control at the controller level can be accomplished using a transfer switch at each panel.

Finally, a manual breaker at each device is how manual control can be accomplished at the device level.

This may sound like something you wouldn't see in this day and age, but you'd be surprised how many systems are operated by a 120-volt circuit breaker that gets manually turned on and off based on the season or time of day.

Automatic Control

Naturally, the opposite of manual control is automatic control.

However, I want to be clear on what automatic control is. I have found that a lot of folks buy into the concept that automating systems is an all or nothing approach. Because of this mistaken belief, some people chose to never attempt automating any of their systems.

I want to clear the air once and for all. This all or nothing belief is not true. This belief has been perpetuated using fear-based marketing to make folks feel like they have to automate everything!

If you take one thing away from this section, it's that you don't need to automate everything, and contractors, do your customers a favor. Let them know that it's ok to only automate a small part of their building. I promise you that you will end up getting more work in the long run.

When it comes to automating systems, there are two approaches to automating building systems. These approaches are:

- Start at the top
- Start at the bottom

Each approach is dependent upon the systems and the technology.

For example, lighting automation tends to do quite well with a bottom-up approach.

This is because you can easily retrofit rooms with occupancy sensors that have spare input/output ports. As your organization matures you can connect these occupancy sensors to your building controls and viola, you are on well on your way towards having an integrated lighting system.

Critical Systems

Every building will have systems that are deemed critical to its operation. These systems typically are:

- Power
- Lighting
- Cooling and heating

Why are critical systems in the control section?

Good question, there is a very specific reason that I put critical systems in the control section.

Have you ever had a power feed or chiller die on you?

If you dig into that situation, you'll often find that there were signs that the failure was going to happen. However, if you don't have control of this system you have no way of detecting and responding to that failure.

Does that mean that control includes monitoring?

From my perspective it does. After all, how are you going to control a system if you can't monitor it?

And if you can't monitor a system how will you be able to maintain it?

Which brings us to the topic of troubleshooting.

Troubleshooting

Troubleshooting is an interesting topic, at some point in your BAS career, you are going to have to troubleshoot a system. With this being such a common task you would think that there would be a common approach to troubleshooting.

Everyone seems to approach troubleshooting differently.

The thing to understand with troubleshooting is that if you don't have a process established then when crap hits the fan you will be less effective and spend much more time trying to resolve issues.

Later in this chapter, I will teach you my troubleshooting process.

As I see it there are three main phases of any troubleshooting process:

- Detecting or being notified of the issue
- Troubleshooting the issue
- Recording your findings

Each of these phases is part of an effective troubleshooting strategy and are critically important to running a facility effectively. If you are weak in one of these areas your entire troubleshooting strategy will suffer.

Maintenance

When I was deciding what order to present these topics in maintenance and troubleshooting were a toss-up. Some would argue that if you "do" maintenance right you will have to do a lot less troubleshooting. I would agree with that in principle but the reason I placed troubleshooting first was that troubleshooting is more difficult than maintenance due to its unpredictable nature.

That is not to short-sell the value of maintenance.

I believe maintenance is so critical, that I have an entire chapter dedicated to it. In that chapter, I will go through a detailed framework for creating a maintenance program and maintaining a BAS.

The takeaway from this section is that maintenance should be looked at as a key function, but it is not the only function.

I've seen too many places where facilities teams look at their jobs as simply performing maintenance and calling it a day. By purchasing this book, you have shown that you have the desire to increase your understanding of how to take a BAS to the next level.

This is why the fourth area, optimization, is an area that I encourage all people to embrace.

Optimization

The term optimization is a term that carries with it a lot of baggage. Ask ten folks what optimization means to them, and you will hear answers ranging from analytics, staffing, and process optimization just to name a few. Because of this teaching how to optimize a building automation system can be a hard subject to unpack.

For the sake of this book when I use the term optimization I am referencing processes and control.

However, I need to discuss a few more things before I dive into my step-by-step processes around troubleshooting and optimization.

Six common tasks that are performed using a BAS

There's a saying that to master any skill you only need to learn six fundamentals. That statement is based on a theory called the Pareto Principle. The Pareto Principle states that 80% of someone's results tend to come from 20% of their actions.

I don't remember exactly where I read this, but I've seen this theory play out enough times in my professional career to make me believe it is true.

When I looked back at the main tasks I did in building and maintaining a BAS there were six tasks that commonly appeared.

These six tasks were:

1. Change settings
2. Schedule systems
3. Add or delete devices and points
4. Create and\or backup databases
5. Perform preventive and corrective maintenance
6. Analyze past performance using alarms and trends

Now hopefully these tasks look familiar to you as I covered them briefly in chapter 2.

Now, that you are much further along in your understanding of BAS I am going to dive deep into each task.

In this section, I am going to cover what the task is, how the task is performed, who performs the task, and what the expected outcome or goal of the task is.

Task #1 Change settings

Changing settings, pretty basic right?

Yep, basic.

However, this is also one of the most service call inducing functions of the BAS.

How you ask?

Think about it, you've got your maintenance guy, let's call him Bob.

Bob is a great guy. He wants to help you save energy and so you've allowed Bob to log into the BAS and turn on the HVAC when he gets into work and to turn off the HVAC when he leaves.

The problem is Bob is not a technician.

One day Bob is in a rush and instead of changing the space temperature set-point he accidently changes a key set-point for your main air handling unit. Bob goes on throughout his day, and when you come in next morning, you realize that Bob's settings have taken down a critical air handler.

Because you didn't consider who your user was, Bob was able to change critical settings.

The lesson from all of this?

Make sure that your user accounts only have access to the points they need to change.

Task #2 Schedule systems

Well, that's enough doom and gloom stories for one chapter, well maybe…

Scheduling, when you want your lights or HVAC to turn on a certain time you use scheduling. Scheduling is super basic, and it should be, after all, you're just turning stuff on and off.

By the way, on and off, those aren't the right terms. In the BAS world, we tend to use occupied and unoccupied.

There is another kind of scheduling you should be aware of, and I like to call this Phil doesn't want to be hot in the morning scheduling. Its official name is optimal start/stop and almost every major BAS supports it.

With **optimal start/stop** you tell the BAS when the building will be occupied and the temperature you want the space to be at. This is awesome because when I want to come in at 5 am and have it be 68°, yes I do start work at 5 am and yes I do like it 68°, I can use optimal start!

Task #3 Add or delete devices and points

Danger, danger!

Adding and deleting points is another key feature of building automation systems that are critical to your success. Eventually, you are going to want to add something or delete something from your BAS.

This task is critically important, but I bet my reason for thinking its critical isn't what you're thinking about.

Sure adding and deleting points is important for replacing controllers or adding sensors.

But do you know why I think you should pay particular attention to this task?

Because this is where your naming conventions and standards go to crap!

There it is, I said it.

You spend all this time creating a great naming standard and Bob, the newly promoted BAS technician, comes in and adds a point without following your standards.

Before you know it you have a hodge-podge of system names that resemble a drunken game of Scrabble.

Task #4 Create and\or backup databases

Opps, do you have a database backup?

Those are seven words you never want to hear one of your technicians ask you. That's why the capability to create, and backup databases are so important.

All of your years of trend data and months of carefully designed systems can be gone with one bad click. If you don't already backup databases, then you need to start.

But, before you do, realize this is a task that requires a dedicated operating procedure.

Matter of fact, if you own or operate a building automation system, stop reading this book.

No seriously, set the book down and write a standard operating procedure for backing up your databases.

You can thank me later.

Task #5 Perform preventive and corrective maintenance

Preventative and corrective maintenance is a task that I could devote a whole chapter on. Oh wait, I did devote a whole chapter to it...

Since you're already here reading this, I figured I'd give you a couple of nuggets of wisdom before you get to that chapter. First off, you perform preventive maintenance to prevent corrective maintenance.

Did you catch that?

Some of the job sites I go to are doing so much corrective maintenance that they never do any preventive maintenance. That's like trying to pay down debt while your spouse is out racking up credit cards, it simply isn't going to work.

That's why you need a solid list of preventive maintenance tasks that your BAS helps you perform. For example, your BAS should be able to alert you when your filter pressure switch triggers or when your cooling coil isn't cooling your discharge air.

These events should be flagged in such a way that they alert the user with not just a high discharge air temperature alarm but with potential causes as well.

Task #6 Analyze past performance using alarms and trends

Analyzing your past performance using trends and alarms is critical to the effective operation of a building automation system. But to do this, you need to understand how to setup and configure your alarms and trends.

Proper setup and configuration will result in actionable data that you can use to drive the right actions.

Let me repeat that because it's critical that you grab onto this concept.

Proper setup and configuration will result in **actionable data** that you can use to drive the **right actions.**

I bolded the parts of this statement that I want to pop out to you.

First off, **proper setup and configuration,** I've been to sites where they trend everything and store nothing.

Really???

Come on!

Are you going to collect all that data and just flush it down the toilet?

This is where a proper setup comes into play. You need to setup your trends and alarms properly.

Next, we have **actionable data.**

Actionable data should be a by-product of proper setup and configuration. If you setup your alarms to alert the right people and you trend the right data, at the right time intervals, then you should get actionable data.

Finally, you are now able to take the **right actions**, rather than just flailing your arms around like a drunken game of pin the tail on the donkey.

Have you ever been at a site where they just respond?

A customer complains that a space is hot or cold and they don't look at a trend report. They don't check to see if there are any alarms. They just walk into the room and randomly troubleshoot.

I want to help you avoid this. If you follow these steps, you can avoid this!

Now let's look at the most underused tool for enabling all the tasks I just described, BAS standards.

How to create effective standards for your BAS

Standards they make the world go round.

Could you imagine if you went to Home Depot and each store called lights something different?

Maybe one store feels like lights should be called glimmers, and another store wants to call them shiners.

Imagine your confusion as you tried to communicate what you wanted to buy.

BAS standards are the same way.

A BAS standard, done right, can tell folks exactly how you want your BAS to work!

You would think that folks would be all over standards. There is a surprising lack of standards across the marketplace. I want to dispel one myth right now. By creating BAS standards, you don't lock yourself to a specific solution.

Now, I'm sure you've got standards and I'm sure their great, but if you want to use a little bit of what I'm teaching here to make them even better go ahead, don't worry I won't tell anyone!

What do BAS standards look like

So what do standards look like?

Surprisingly different.

There is no "official" standard, for standards. Sure some folks will argue that the Construction Specification Institute Master Format is a standard. It is, but it is not a BAS standard, and quite honestly it won't get you what you want.

So what should standards look like?

Well, I will tell you this. Standards shouldn't be massive, hulking documents that force mere mortals to cower before your BAS mastery. No, standards should be light, agile and able to adjust, while at the same time being rigid enough to keep your BAS system in order.

What should BAS standards include

A BAS standard should include the following six areas:

- System Profiles
  - Naming
  - Trend Intervals
  - Alarm Settings
  - Graphics
  - Device Specifications
  - Controller Wiring

I would not feel comfortable endorsing a standard that was missing any one of these categories. However, I also know that we all need to begin somewhere and that somewhere might be a very different place for two different people.

Therefore, don't give up, even if you get 10% of what I describe below, you are still 10% better than you were before.

So, what should a standard look like?

Well, you are in luck! I'm going to build one with you right now.

Our scenario

In our scenario Facility Manager Bob, this guy just keeps getting promoted, has decided that he shouldn't be doing BAS projects ad hoc.

Therefore, Bob realizes he needs to come up with a standard. Bob figures the first thing he needs to do is to create a common naming standard so that everyone is speaking the same language.

In this scenario, we are going to build out a system profile for a VAV box.

System profiles

System profiles are just that they are profiles of a system. A system profile will enable you to create standards based on system types. This will make it so much easier for your service providers and internal staff to check something is "in compliance" with its standard.

Naming

Oh, naming where to even begin with you!

When it comes to field controllers, I recommend that you use a combination of the space name and the equipment name.

Then I recommend that you create a list of what points each device type will have.

Naming standards

Field Controllers will be named based on their system and location. An example of this would be:

- RM 205-VAV

A reheat VAV box will have the following points named and exposed:

- ZN-T: Zone temperature
- DA-T: Discharge air temperature
- ZNT-SP: Zone temperature set-point
- DPR-O: Damper output command
- CLGCFM-SP: Cooling cubic feet per minute set-point
- MAXCLGCFM-SP: Max cooling cubic feet per minute set-point
- HTGCMF-SP: Heating cubic feet per minute set-point
- HTG-O: Heating output

Trend intervals

Next, you need to define your trend intervals.

Now as I mentioned in Chapter 2, there are two types of trends, CoV and time interval. Remember that CoV, stands for Change of Value and the range is the full range of the point.

So, a 0 – 100% output would have a range of 100. 1% of 100 is 1.

Now that we have our naming standards we are solid! Let's continue by building out our trend intervals.

Point Name	Point Description	Trend Interval
ZN-T	Zone Temperature	300 Seconds
DA-T	Discharge Air Temperature	300 Seconds
ZNT-SP	Zone Temperature Set-point	300 Seconds
DPR-O	Damper Output Command	CoV (1% of range)
CLGCFM-SP	Cooling CFM Set-point	CoV
MAXCLGCFM-SP	Max Cooling CFM Set-point	CoV
HTGCMF-SP	Heating CFM Set-point	CoV
HTG-O:	Heating Output	CoV (1% of range)

Alarm settings

Next, the alarm settings need to be determined for each of the points that have been defined for the VAV reheat unit. This is done by listing out the high and low settings for each alarm point.

Now one of the important things to note is that you only want to list alarm settings on points that can go out of range. Set points and outputs don't normally go out of range so you wouldn't set alarms on them.

Point Name	Point Description	Trend Interval	Alarm Hi	Alarm Low
ZN-T	Zone Temperature	300 Seconds	85°	55°
DA-T	Discharge Air Temperature	300 Seconds	160°	70°
ZNT-SP	Zone Temperature Set-point	300 Seconds	N/A	N/A
DPR-O	Damper Output Command	CoV (1% of range	N/A	N/A
CLGCFM-SP	Cooling CFM Set-point	CoV	N/A	N/A
MAXCLGCFM-SP	Max Cooling CFM Set-point	CoV	N/A	N/A
HTGCMF-SP	Heating CFM Set-point	CoV	N/A	N/A
HTG-O:	Heating Output	CoV 1% of range	N/A	N/A

Graphics

Everyone loves graphics but very few people want to put time into developing them. Did you know that even if you just drew your graphics template on a piece of paper, you would still be better off than 90% of the standards I've seen?

Most standards include what points should be on the graphic but beyond that? Forget about it!

Graphics are often left up to the designers, and the designer's idea of a good graphic might be very far from your own.

That is why you need to detail out your graphics.

Draw your graphic in MS Paint or simply sketch a mock-up of your graphic and scan it into your specifications. This is an important step that so many people miss!

For our VAV box, we would want to have a space type graphic.

Device specifications

In some of the standards, I've seen folks will go as far as to specify the temperature range of the controller, its enclosure type, power source, etc.

I'm going to be a bit of a dissenter and say that, in most cases, specifying those details is not necessary. Sure there are situations that justify specifying controller build quality, but those are usually very limited.

In reality, the area you should focus on are the capabilities.

Do you want the device to have an onboard display?

Does the device need a time-clock so it can run independently of the supervisory device?

Do you want the device to be capable of supporting Power over Ethernet?

These are all requirements that should be put under the device specifications.

In the case of our VAV box example, we will state the following.

The VAV box controller shall have the following capabilities:

- A proportionally driven actuator that is part of the controller assembly.
- 2 Universal Inputs and 32 Universal Outputs
- 24VAC power
- An Actuator that can drive its full range in 90 seconds or less
- On-board Air-Flow sensing capabilities

Controller wiring

Controller wiring?

Why would you create a standard around that?

Have you ever tried to go and troubleshoot a 10-year-old installation?

Sifting through dozens of white wires to find which wire goes to which device is a PAIN!

This can be fixed oh so easily. I recommend that you create a wiring standard for your controllers.

Here is my recommended setup:

- Analog Input- White Wire
- Binary Input – Yellow Wire
- Analog Output- Orange Wire
- Binary Output- Red Wire
- Power – Black Wire
- Communications Bus – Blue Wire

Putting it all together

So, how do you put all this together?

Where do you start and how do you go from having no standard to having a standard your mama would be proud of?

Well, you're in luck, I've detailed out a step-by-step process in the next section.

The process

There are four steps for the standard creation process. These steps are:

- Identify what you have
- Sort through what you have
- Identify where you want to be
- Take action on the gaps

Identify what you have

In this step, you are going to go through all of your past documentation. You are looking for any word documents or PDF documents that contain specifications or standards. During this process, you are solely focused on gathering data.

You can store all of this information on a local or cloud storage service (DropBox, OneDrive, etc.)

Sort through what you have

Remember those six areas I covered early in this Chapter? If you don't here, they are again:

- Naming
- Trend intervals
- Alarm settings
- Graphics
- Controller wiring
- Device specifications

Now you are going to go through each of these documents and look for any standards you have. Then you will pull these standards into a Word document. This Word document is going to have six categories in it.

Can you guess what those six categories will be?

Yep, they will be the six categories from the earlier section.

At this point in the process, this is a copy and paste exercise. If you think you could use it copy it into the category, you believe it belongs in.

Rather quickly you are going to realize where your existing standards are strong and where they are weak. Now it's time to make a decision…

Identify where you want to be

You need to make a decision on where you want to be with your standards.

Are most of your problems due to poor alarm management?

Or do you run into issues with inconsistent naming?

Here you will determine where you want to be and you will note those areas as gap areas.

Take action on the gaps

In this step, you will begin to build out each category based on the method I showed you earlier. So, if you recognized that you had an issue with alarm standards, then you will follow the process I detailed out earlier on how to write out alarm standards.

My step-by-step troubleshooting process

I like to think that one of my strengths is my ability to troubleshoot. For some reason, this is an area that comes quite naturally to me. I have a 9 step process that I like to follow when I am approaching troubleshooting problems.

To illustrate these nine steps, I am going to walk you through a couple of the more memorable problems that I've troubleshot. The first example I'm going to run you through is a fairly unique problem I discovered the other day working on an integration project. The situation was that I was trying to get a camera feed to pop up to display on a screen when a person tried to use their access card to access a restricted space.

The problem was that the video feed would not display.

Step 1: Identify the symptoms

First I identified the symptoms of the problem. In this scenario, I noticed that the video feed would not display and I was getting an error saying that the video server could not be accessed.

I knew from experience that this error meant that there was a DNS issue. I went to the host file and resolved the IP address to the domain name, and the issue was fixed.

Now I am going to go through another situation that I encountered. In this situation, I had a touchscreen that would not work.

Step 2: Google the problem

I had a touch screen that wouldn't work. It had a window that popped up with an error code. I Googled the error code and found that I was supposed to have a driver installed. I installed the driver, but the touch screen still wouldn't work.

Step 3: If it worked, find what changed

This was a brand new installation, so I had to look at an existing touch screen to see how those were setup. I was able to find a setting that was different, but I couldn't figure out how that different setting was installed.

Step 4: Roll back to recent state if possible

Fortunately, I was able to look at the actions that had been performed on the machine connected to the touch screen monitor. From here I was able to figure out that a different driver had been installed. I installed the other driver, and now the touchscreen worked...

Step 5: List out the probable causes

If you have multiple things that could be causing an issue, it helps to list them out. There was a time I was working on a BAS that was connected to a data center. By the time I got brought in the data center had been dropping communication with the BAS every night.

After trying to resolve the issue unsuccessfully, I started to list out the issues.

Step 6: Write out your actions and expected results

With this list of issues, I started to write out the actions I could take and what result I expected to have. The first action I could take was to capture five nights worth of network traffic. My expected result was to identify the packets that were crossing the wire when the BAS went offline.

Step 7: Perform the task

I performed this network capture, and I was able to identify that a SQL database backup was taking place across the WAN link. This was causing the BAS traffic to get dropped resulting in the BAS going offline.

Step 8: Repeat steps 5-7

If I had not found the issue, I would continue steps 5 through 7 until I did.

How I optimize a BAS

One of the most common tasks I perform is helping folks optimize their building. When I first visit a building, I perform 12 checks on the BAS. Once I have performed these checks I provide the customer with a report on the actions they should take.

Here are the 12 checks that I perform:

I will be discussing these tasks in a future video series.
Check 1: Check out the schedules
Check 2: Look for overrides
Check 3: Prioritize the alarms
Check 4: Create resets
Check 5: Sync the schedules
Check 6: Tie back to access
Check 7: Tune loops
Check 8: Calibrate inputs and outputs
Check 9: Analyze the trends
Check 10: Create resets
Check 11: Implement demand ventilation
Check 12: Setup regular reports for trends and alarms.

Chapter 10 quick summary

Yes!

I don't know about you, but I feel like I packed enough value bombs in this chapter to finish the book right here (don't worry I've still got a ton of information left to cover!)

In this chapter, I took you through the four main ways facility teams use BAS systems. I hope that my exploration of these topics was eye opening for those of you who have never worked in the service world.

I wanted you to come away with an appreciation for what customers have to deal with on a daily basis. I feel that a lot of BAS designers live in a vacuum and design BAS based solely on the scope without considering how a system will be used.

That's why I spent a big chunk of this chapter laying out the common tasks that are performed by owners and service professionals.

Next, I went and talked you through how to create standards for your building automation system. This is a critical topic, and I encourage you, if you are an owner or service professional, to create a BAS standard right now!

Don't wait any longer!

Finally, I closed out the chapter by giving you a step-by-step process for troubleshooting and for optimizing your building automation systems.

Get ready because in the next chapter I am going to be teaching you how to manage maintenance tasks and the service providers who perform them!

CHAPTER 11

Managing Maintenance Tasks and Service Providers

What's in this chapter

Managing maintenance tasks and service providers is a critical task but what does this have to do with BAS?

The reason I am including this chapter is that a BAS can and does have a massive impact on how you perform your maintenance tasks and how you select service providers.

In chapter 10 I covered the common tasks and functions related to a BAS from a facility manager's perspective. In this chapter, I am going to look at how to service a BAS whether you are a building owner or a service provider.

I started my career as a service technician. It was in this role that my experience and knowledge of BAS skyrocketed. The reason I learned so fast is that servicing a BAS is an extremely complex task.

In this chapter you will learn:

- The different approaches to servicing a BAS
- How to set up an effective maintenance program
- Key maintenance tasks for your BAS

The different approaches to servicing a BAS

In the world of BAS, you have two primary approaches to servicing a BAS. The first approach is what I like to call in-house service. **In-house service** is where you or your staff execute all of the tasks associated with servicing your building automation system.

The second approach is called planned service. **Planned service** is where you utilize an outside party, typically known as a service provider, to provide you planned service for your building automation system.

Neither of these approaches are necessarily better than the other. Rather it depends on your specific building needs.

In this section, I'm going to discuss in-house versus planned services, and I am going to show you where each approach would apply to a building. The first thing I am going to cover is in-house services

In-house

In-house services are typically executed by a facility department or group.

This group will typically consist of a leader known as a facility director or facility manager. This group will typically be divided into mechanical, electrical and controls specialists. These specialists will have a defined set of preventative maintenance tasks.

You will tend to want to perform in-house maintenance if you have an experienced BAS control staff or a relatively simple building automation system. Either of these scenarios tends to respond well to an in-house services group.

Types of maintenance

There are three types of maintenance tasks that can be performed those are:

- Preventative maintenance
- Reactive maintenance
- Just-in-time maintenance

Preventative maintenance consists of tasks designed to maintain your building automation system. These tasks consist of various items which I will describe in the "how to set up an effective maintenance program" section.

Reactive maintenance is the opposite of preventative maintenance. With reactive maintenance, you are reacting to failures in the system. These are often known as trouble calls and are the bane of most facility departments.

The goal of the facility group, in regards to maintenance, is to minimize the amount of reactive maintenance while continuing to perform preventative maintenance.

There's a third type of maintenance that has come up recently, and this is known as just-in-time maintenance. **Just-in-time maintenance** uses analytics or advanced functions of the building automation system, to enable the facility staff to perform maintenance based on the real-time status of their building automation system. This allows the facilities group to avoid or reduce the amount of preventative maintenance they need to perform.

Planned service agreements

Planned service agreements or PSA's, break out into three different strategies.

These three strategies are:

- Full coverage
- System specific
- Block hours

Before I dive into these three strategies, I wanted to take a second to talk through planned service.

I started my career performing service on building automation systems. I loved this. Every day was different, and I never knew what I was going to walk into. But on the flip side for building owner, that unpredictability can be overwhelming and expensive.

Additionally, if you or your group are inexperienced with building automation systems, it may be difficult for you to know where to begin. To deal with this, service provider organizations have developed processes and programs to help building owners service their building automation systems.

My goal here is not to try to sell you on utilizing service providers. I simply want you to understand the capabilities that exist and what the different levels of service mean to you or your customers.

Full Coverage

Full coverage is the first form of service that I am going to discuss. It is also the most expensive form of service.

In a **full coverage maintenance** scenario, the service provider is going to evaluate the systems that exist and provide a price to perform full coverage maintenance of these systems. This form of service tends to work best in buildings where the building automation system is relatively new and was provided by the service provider.

This is because new equipment tends to fail less. Also, the service provider tends to understand their building automation system best.

As part of a full coverage agreement, the service provider will define the tasks that they will perform. This is similar to a system specific agreement, which I will cover next. The service provider will perform all maintenance and troubleshooting related to the systems they are covering.

Typically, the service provider will do this by putting a BAS technician from their staff on-site. In some buildings, this will be a 24-7 position and will require multiple people.

This is the most expensive form of service because of the level of service provided.

System specific

System specific service is where the service provider will perform service on specific systems.

For example, the service provider may go and only service the main supervisory controllers for the building automation system and the building automation controls for the larger systems (e.g. chilled water plants and air handlers).

As part of a system specific service contract, the service provider will list out a set of tasks that they will perform on a regular basis. The service provider will also provide a list of reports that will be run to show the status of the systems they are maintaining.

This form of service tends to work well for organizations that have their own in-house building automation teams but do not have the manpower or skill level to service the larger, more advanced systems.

The price for this kind of coverage depends on the systems age, size, and complexity.

Block hours

Block hours are the final form of planned service agreements. In **block hours** agreements the owner is purchasing a block of hours for a specific level of technician. The owner can then utilize this block of hours to perform whatever task they would like.

This form of service tends to work well for experienced organizations that would like to augment their staff on occasion. It is critical that the skill level of the technician who will be working for the customer is defined before pricing this kind of agreement.

You might have a problem if you get the wrong kind of technician for your block of hours. What if you intended to use these hours for maintaining a central plant controller and the technician that the service provider priced only has the skill level to work on VAV boxes?

To avoid this, you need to be clear with the service provider what potential tasks you are looking to use the block of hours for.

How to set up an effective maintenance program

In this section, I'm going to discuss how to set up an effective maintenance program. I want to be clear that I am not laying out a maintenance program for HVAC systems. This maintenance program is focused on building automation systems only.

The funny thing is, in my experience, this is where folks mess things up. It's not due to lack of knowledge it's simply due to there not being a clear set of tasks and processes related to building automation systems.

In this section, I'm going to lay out a clear step-by-step process that you can follow to identify the tasks you need to take to maintain your building automation system.

Specifically, this section will cover:

- How to identify the systems you have
- How to prioritize your systems
- How to identify the tasks for your systems

At the end of this section, I will provide you a sample of a set of maintenance tasks for a building automation system.

How to identify the systems you have

At first, I wanted to tell you that identifying the systems in a building is quite simple. I wanted to paint a picture for you that you could login into a building automation system run a report and discover the systems that currently exist.

However, I know from experience that this is not always possible.

So how are you going to be able to identify the systems that exist throughout a building?

You can do this by using the process I am going to lay out. If you follow these steps, you will know what is in a building. This process will work for you whether you are an owner or a contractor.

Step One: Identify the vendors you have

You need to identify what vendors you have within your building. Not only will this be critical for future support and troubleshooting but it's also going enable you to offset a lot of the costs associated with discovering what's in your building.

Step Two: Contact the vendors whose systems are in your building

Why do all the work yourself, whether you're an owner, contractor, or an engineer you can reach out to the organizations who provided the systems for the building you're working on? Quite often this will help you shortcut the data collection process.

Step Three: Dig through the documentation

I wish I could say something a little sexier than dig through the documentation but in reality, that's what you have to do. You are going to have to go through the drawings of the building and identify the systems that exist.

As you go through the drawings you will list out the systems according to the following information:

- BAS manufacturer
- Controller Location
- Controller name
- Function of controller
- Drawing number reference

Now to get some of this of information, you may have to go into the program of the controller itself. For those of you who already have this information and you are sure it is accurate you can skip this step.

Step Four: Gather whatever documentation you do not have

In this step, you are going to make sure that you didn't miss anything.

How are you going to do that?

Well, I have a couple of different ways.

First, you are going to ask around.

Find the contractors who have worked on your projects in the past. Contact your current employees who may have documentation related to your building automation system.

Now I realize that this seems basic, but you be surprised, when folks get busy they tend to forget simple tasks.

Next, you are going to walk to each of the mechanical rooms and validate that the information you collected accurately represents what is installed.

Finally, you are going to take one last look at the building automation system through its graphical user interface, to see if you missed anything.

Step Five: Consolidate the information

In this step, you're going to take all of the information you collected in steps one through four, and you're going to consolidate that into a document.

This document should have the following information inside it:

- Building automation system
  - System version
  - System installer
  - Location of the main server
  - Server credentials
  - Copy of the server database
- Building automation supervisory devices
  - Location of the supervisory devices
  - Current version of software
  - IP address
  - Domain name
  - Supervisory devices credentials
  - Copy of the database for the supervisory devices
- Building automation field controllers
  - Location of the field controllers

- Current version
- Network and or Fieldbus address
- Controller name
- Controller function
- Program inside the controller
- Controls drawing and sequence of operations

Upon completing this document, you will have a solid list what inside your building and how it functions.

Now you are going to move on to the next task, which is prioritizing your systems.

How to prioritize your systems

In the previous section, you identified what systems you have in your building.

Now in the real world, all systems are **not** created equal. Although it may be fun, to spend 10 hours troubleshooting a VAV box in the lobby, it is not as effective as making sure the PI loops in a central plant controller stay tuned.

That is why **you must** do this task.

Do not, do not, **do not skip this task**.

Okay, I think I got my point across

To make sure you stay focused on maintaining the systems that are critical to your building's performance, you need to rank the priority of the systems. So how do you do that?

It's rather simple. Below I have given you a priority ranking matrix.

Criticality Matrix	Work Stoppage	Regulatory Compliance	Life-Safety
Required (5)			
Significant (4)			
Moderate (3)			
Minor (2)			
Low (1)			
Overall Ranking			

I laid out five levels of criticality, in the matrix above. These levels range from required to low. What each of these levels represents is the impact to the three categories on the right of the criticality ranking.

For each of the three categories, you will choose one level of criticality. On the bottom row of the matrix, you will put a number for the overall ranking. This is simply the number on the right of the criticality level you selected.

You will want to look at each system type through this matrix. Now I want to be clear, what I mean by every system is not each VAV box. Rather, you should look at each system type and rank those accordingly.

The matrix below has been filled out, for a central cooling plant at a hospital.

Criticality Matrix	Work Stoppage	Regulatory Compliance	Life-Safety
Required (5)	X	X	X
Significant (4)			
Moderate (3)			
Minor (2)			
Low (1)			
Overall Ranking	5	5	5

As you can see, the central cooling plant ranked required in each category. This was due to the effect that a loss of cooling would have at the hospital.

This is a quick and easy way to rank your systems. I'm going to take a brief moment to unpack each of the ranking categories means.

Work stoppage

Work stoppage is another way of saying that an impact to the system will keep the facility from performing its main task. In the case of a hospital, a lack of cooling will make the hospital unable to perform certain functions.

Regulatory compliance

Regulatory compliance deals with the rules and regulations that the facility is responsible for adhering to. In the case of hospitals, certain regulatory bodies require hospitals to maintain specific environmental conditions to take care of patients.

Life safety

Life safety has to do with keeping the occupants within a facility safe. A loss of cooling would have a detrimental effect on the safety of patients and personnel within the hospital.

As you can see in some ways, these categories can be subjective. The purpose is not perfection, simply progress. Too often systems are lumped together and are not prioritized.

Simply by prioritizing your systems, you will be much further ahead than others.

How to identify the tasks for your systems

How can you identify the tasks that you should perform for your systems?

What systems should you perform these tasks on?

In this section. I'm going to describe the maintenance tasks that should be performed for a BAS. These are the tasks that should be performed to maintain a building automation system at peak performance.

The systems I will cover will are:

- Servers
- Database
- Supervisory devices
- Controllers
- Sensors
- Relays
- Actuators
- Analog outputs

Servers

To properly maintain a building automation server, you need to perform a select set of maintenance tasks. These are the tasks you need to perform and the frequency you need to perform them:

Task	Description	Frequency
Check the availability of server resources	In this task you will be ensuring that your server has adequate storage, memory, and CPU. You will do this through the use of logs	Once every three months or anytime the major changes made
Backup image of server	In this task you will create a backup of the server and store it external to the server. This is not the same as creating a backup of the database	Once every three months or anytime a major change is made

Database

The database is where all of your trends, alarms, user preferences or data and system-level backups exist. As you would imagine this is a critical system for you to maintain.

Now I realize that most of the folks reading this book are not database administrators. When you see these tasks, realize you'll be working with the IT folks identified in chapter 4.

Task	Description	Frequency
Backup the database	In this task you will be ensuring that your database is backed up to an external storage device.	The frequency of a backup is determined by how much data loss is acceptable. In most large organizations databases are backed up nightly.
Purge old records from the database	In this task you will be removing old historical records from the database and transferring them to some form of external storage.	The frequency of this task is determined by how much data you want to retain in your database. Typically, organizations will remove historical data that is five years or older.

Supervisory devices

Supervisory devices, as described in Chapter 2 provide overall management of building automation controllers, communication trunks, and network-level logic.

Task	Description	Frequency
Check the availability of supervisory devices resources	This is a highly important yet often missed task. In this task you are evaluating your supervisory device to ensure that it has adequate resources available.	The frequency of this task depends on how often devices and/or points are added to the supervisory device. Typically, this task should be done monthly.
Backup image of the supervisory device	In this task you will create a backup of the supervisory device. This is not the same as creating a backup of the database.	Once every three months or anytime a major change is made
Check on the health of communication trunks	In this task you are making sure that your communication trunks are healthy. You're also verifying that your controllers are able to communicate with the supervisory device.	This task should be performed every week and anytime a controller is added to the communication.

Controllers

Controllers make the BAS world go round. Without controllers nothing happens. Controllers allow you to control the HVAC and other systems within your building. You would be shocked by how many sites I visit that have controllers off-line and folks are acting like it's not a big deal!

Task	Description	Frequency
Check the connectivity of your controllers	In this task you will be ensuring that all controllers are online	This task should be performed every week and anytime a controller is added to the communication
Backup the program within your controller	In this task you will be creating a backup of the program within your controller. This backup will be stored on an external device	This task should be performed any time a controller program is changed and/or a controller is tuned
Tune your control loops and control blocks	In this task you will be tuning the control loops and control blocks within your controller. Specifically, you will be tuning PI loops, sequencers, and timers	This task be performed on a semiannual basis or anytime a program is changed within a controller

Sensors

Sensors act as the inputs to your controllers. There are a variety of sensor types ranging from temperature sensors, pressure sensors, and binary sensors (status switches and safeties)

Task	Description	Frequency
Verify the accuracy of your sensors	In this task you ensure that your sensors are accurate. You'll do this by adjusting the offset within your controller to match the measured value of the sensor	This task should be done on a semi-annual basis or anytime a sensor is replaced/installed
Ensure the operation of your binary sensors	In this task you'll be testing your binary sensors to ensure that they are operating correctly.	This task should be done on a semi-annual basis or anytime a sensor is replaced/installed

Relays

Relays allow you to control devices using an on-off signal. A relay allows you to pass voltage to a set of contacts versus having to source it from your controllers.

Task	Description	Frequency
Ensure that the relay works	In this task you will be ensuring that the relay coil can be energized and that the contacts open and close	This task should be done yearly or anytime a new relay is installed

Actuators

Actuators are devices that are used to make valves and dampers move. Actuators utilize a variety of different signals for their operation.

Task	Description	Frequency
Ensure that the actuator works	In this task you will check that the actuator drives fully open and fully closed. And if the actuator is a proportional or floating actuator you will want to check that the actuator is able to accurately meet the following positions 25% 50% and 75%	This task should be done yearly or anytime a new actuator is installed

Analog Outputs

Analog outputs are used to drive a variety of devices like, VFD's, sequencers and other control devices. Ensuring that you have the proper signal both leaving the controller and at the controlled device is important.

Task	Description	Frequency
Ensure that your analog output signal matches the output of your controller	In this task you will be ensuring that the analog outputs controllers are providing the correct will to in that you are detecting the correct voltage at controlled device.	This task should be done yearly or anytime a new device and/or controller is installed

How to Apply this Knowledge Today

Chapter 11 Quick Summary

I covered a lot of information in this chapter.

In this chapter, I showed you the different ways a building automation system could be serviced. As I did this, I showed you what the differences between in-house and planned service agreements are. Next, I laid out when you may want to use these different approaches.

From there I went through the process of setting up an effective maintenance program and system prioritization. By the way, this process may look similar to the system matrix I showed you in Chapter 10. That's awesome if you picked that up. I want you to get comfortable with using the system matrix to prioritize systems

Finally, I wrapped up this chapter by providing you a list of tasks based on the different parts of a building automation system. In each table, I listed out a set of tasks, a description of these tasks, and the frequency at which you should be performing these tasks.

Before I leave this chapter, I want to call on you to apply what you've learned. Whether you are a contractor, engineer or owner, you can still apply this knowledge right now!

In the next chapter, I am going to dive deeper into BAS management! You ready?

Let's go.

CHAPTER 12

ADVANCED MAINTENANCE MANAGEMENT

What's in this chapter

What?

Another chapter on maintenance programs?

Didn't I just cover this in chapter 11?

If you think that, don't worry, while chapter 11 was very much a hands-on chapter. This chapter is going to look at the higher level aspects of maintenance.

In this chapter I'm going to discuss:

- The financial aspects of maintenance management
- How to evaluate the ROI of your actions
- How to retain and grow the tacit knowledge within your organization

As you will see, this chapter is more focused on the financial and operational questions related to maintenance.

The financial aspects of maintenance management

Believe it or not, it is about the money.

At the end of the day, many things don't get done because there is not enough room in the budget to pay for all the tasks that need to happen. Unfortunately, when it comes to budgeting, building automation systems are often the budget's forgotten step children.

After all, it's quite easy to justify maintenance on a big piece of machinery or some form of interior decoration that is needed to cover up a glaring eyesore. However, investing in a new supervisory device because your older devices are running out of space or are out of date, is very hard to visualize.

That's why in this section I wanted to unpack the financial aspects of maintenance management.

Depending on whether you work in the public or private sector your funding process may look a little different. However, one thing that is common across the industry is that budgets breakout into capital and operational budgets.

Each year, customers evaluate the current cash flow they have within their organization and allocate certain amounts of money for the different departments to utilize. This money is typically divided between capital and operational budgets.

Each year the leader of the facility group will solicit a list of maintenance tasks that need to be done from his or her team. This list of tasks is then prioritized based on a variety of factors.

The most common factors are:

- Return on investment (ROI)
- Impact to the business

Return on Investment

Return on Investment is a form of measurement used to determine the amount of time it will take an investment to produce a payback. On paper, this can be a great way to prioritize operational investments.

Things like equipment replacements tend to lend themselves well to a return on investment evaluation. Unfortunately, operational investments into the building automation system tend not to align well with a return on investment measurement.

Which brings me to, the measurement that does seem to matter, the impact on the business.

Impact to the business

The impact that the degradation or failure of the building automation system can have to a business is an area that can be quantified. That is why this is one of the areas that I often recommend folks look at when they are receiving pushback on the inability to define an ROI for a building automation system maintenance activity.

How would you measure the impact to the business?

If you recall, in Chapter 10, I laid out a criticality matrix.

You can utilize this matrix, to determine the financial impact to the business. Once this matrix is filled out you can use the potential business impact to help you position the value of any maintenance activities you want to pursue.

Criticality Matrix	Work Stoppage	Regulatory Compliance	Life-Safety
Required (5)	X	X	X
Significant (4)			
Moderate (3)			
Minor (2)			
Low (1)			
Overall Ranking	5	5	5

As you may recall this matrix represented the central cooling plant for a hospital.

How could you take this matrix, and use it to show the financial impact of a system failure?

Here is how I would approach this problem.

First, you would need to gain agreement, that the loss of a central plant controller would indeed cause a work stoppage, an impact to regulatory compliance, or a risk to life safety. You can achieve this by explaining what the failure of a plant controller would mean to the hospital.

Once you've gained agreement that the loss of a central plant controller can cause this scenario, you can begin to explain your needs. You can explain how the central plant controller is at risk and requires a maintenance investment.

For example, let's say that your central plant controller is 20 years old and can no longer be purchased. The risk to the business is that you may not be able to find a new controller or someone to work on the controller if it fails.

Next, you would go and define how long it would take to find a new controller and someone to install it based on experience. You would then detail out the cost of manning the central plant 24/7 to maintain plant operations.

This cost, along with the potential cost of losing the ability to use certain parts of the building, would be used to justify this investment.

How to evaluate the ROI of your actions

Sometimes you're going to be asked for an ROI no matter what.

In cases where the building automation system has not been maintained or upgraded for long periods of time, you may be able to implement some controls strategies and create a real return on investment.

An example of this would be, a building with an older pneumatic system that is operated via a time clock and has no method of providing feedback to the user. The problem with this scenario is that it would most likely require a capital event to replace this system.

So then how can you show an ROI for maintenance activities?

It almost feels like I am going in circles and am trying to fit a square peg into a round hole.

Fortunately, the task of creating an ROI for a maintenance activity is not impossible. It's just quite difficult.

If your manager or your customer insists on you providing a return on investment here is how you can address that.

Calculating the ROI

First I should detail out exactly what an ROI is.

To calculate an ROI, you take the gain from the investment minus the cost of your investment and divide that by cost of the investment.

$$ROI = \frac{(Gain\ from\ Investment - Cost\ of\ Investment)}{Cost\ of\ Investment}$$

As you can see in the equation above, the gain from the investment minus the cost investment is divided by the cost of the investment and that equals the ROI.

Your **breakeven** point is when your cash flow turns positive for your investment. Meaning that you have paid back the initial investment and your investment is now saving you more money than you are spending on it.

Retaining knowledge within the organization

So far throughout this book I've talked a lot about the importance of maintaining the documentation and information surrounding a building automation system.

Now it's time to talk about how to maintain the knowledge that resides within the facilities team. This critical topic is often not discussed, and to date, I've yet to see any process laid out to help folks understand how to collect, store and transfer the knowledge within an organization.

In this section, I'm going to cover just that. Whether you are a building owner, contractor, or an engineer, you can utilize this information to retain and transfer knowledge.

When I started this section, I figured I'd come up with some witty acronym, but nothing came to me, so you are stuck with, CST. CST does not stand for Central Standard Time. CST stands for collect, store and transfer. It is through this process that information will be collected, stored and shared with other personnel.

Collect

Collecting information is often the hardest part of retaining knowledge especially when that information is within someone's head.

Unfortunately, you are not in the movie The Matrix, and you can't just jack into someone's head and pull out all their knowledge, and honestly with some folks would you even want to be in their heads?

So how can you take the historical knowledge that remains within your experienced team member's heads and transfer it to the rest of the organization?

Over the years I've tried a lot of things to accomplish this. I've tried working sessions, having more experienced technicians mentor younger technicians, and I've even tried interviews over coffee and donuts.

However, nothing seems to work as well as the good ole shadow approach. In this approach, you simply follow a technician and document the process that he or she uses to complete a certain action.

Some of you may be asking, "Why I would want to go through all that trouble?"

The thing is there is a uniqueness to most building automation systems. Often there's no documentation that details out how to service a piece of equipment that was custom built or is no longer supported.

The only way you can capture this information is by sitting down and documenting how it is done.

But what do you document?

What information do you collect?

If you recall earlier in chapter 11, I listed out a series of tasks that you should perform for each of the systems you have. I propose that a good place to start would be to document how each one of those tasks is performed for your specific systems.

Now, what would this look like in the real world?

Because I don't want you to think I'm just telling you to do things that aren't done out in the field. I am going to tell you about the way I have done it.

I tend to document the procedures that need to be performed by listing them out on a laminated card.

These cards are often attached to a piece of equipment or control system using zip ties or some cord. Here is a step-by-step process to create these process cards.

Step 1: Identify the systems detailed out in chapter 11.

Step 2: Identify and agree upon the processes and tasks that your team regularly performs on the systems.

Step 3: Find someone on your team or hire a service provider who is an expert in performing these tasks.

Step 4: Work with this expert to document each step of the task using text or pictures where appropriate.

Step 5: Store this documentation in a local or cloud storage environment (more about this in the next section).

Step 6: Print out this documentation and attach it to the system on which these tasks are being performed.

Step 7: Additionally, print out this documentation and store it in a book in a centralized location.

Storing information

Okay so you've went, and you've captured all this information, what you do now?

Well the first thing you need to do, is you need to find a place to store this information. Where you store this information depends on your local environment and how technically savvy you are.

I recommend that you store your documentation on some form of cloud storage like Google Drive, Box or Dropbox.

So how do you do this?

In the following steps, I detail out how to index and store the information you gathered.

If you followed along in chapter 11 and created a list of the systems and the tasks you need to perform then you're already ahead of the curve.

Step 1: In this step, you will create folders within your online or local storage account for each of the systems that you're performing tasks on.

Step 2: Next you will take the tasks for each system, name them, and add them to the appropriate folder.

Step 3: Finally, you will create a document for each folder that lists out each task and a description of the task. This will allow anyone who is accessing the folder to understand what tasks are in that folder.

Transfer

Now for the hard part. You need to figure out a way to transfer this information to new and future employees.

How can you do this?

Fortunately, I've got a step-by-step process for this as well!

In the following steps, I will show you how to transfer information from your knowledge database to your new and future employees.

Step 1: Segment your company's employees into different skill levels. Determine the tasks that each level will need to perform.

Step 2: Detail out when you will expect these employees to perform these tasks.

Step 3: Create a document that states, the employee's role and the tasks you expect them to perform.

Step 4: Create a checklist that contains the employee's name, role, and tasks.

Step 5: Have the employee perform the task in front of a senior employee. Have that senior employee sign off that the task was completed properly. For complex tasks, you may want to have the senior employee observe the junior employee perform the task multiple times.

Step 6: Have employees regularly recertify on the task(s) they are expected to perform.

Chapter 12 Quick Summary

In this chapter, I walked you through some of the managerial aspects of maintenance. As I said earlier in the book, I spent a good chunk of my life providing service to customers with BAS.

When you have to create work to survive, you learn to sell the value of your activities. But here's the deal…

What you sell cannot be a load of BS. The BAS world is small, and you will get called out real fast if what you are providing is a ball of crap. That's why in this chapter I taught you how to show the real value of the maintenance activities you engage in.

I did this by walking you through the financial aspects of maintenance management. I taught you how to show the financial ramifications of not maintaining a system and how to prioritize your activities based on ROI.

I then taught you how to collect, document, and retain the knowledge that exists within an organization. **This is critical information!**

Seriously, there are only so many of us BAS folks to go around, and I'm sure you've seen that without this book, it's pretty darn hard to find educational resources.

That's why it's critical that you begin to capture and document the knowledge within your organization!

Do this today!

In the next section, I get to the most anticipated part of the book. I know the phrase "I've saved the best for last" is cliché, but in this case it's true! In the next section, I'm going to teach you about analytics, IoT, and systems integration.

Buckle up because I'm about to take you on a ride!

SECTION IV – ADVANCED TOPICS ANALYTICS, IOT, AND INTEGRATION

Section Overview

It's time for the training wheels to come off.

You're no longer constrained to the fundamentals of building automation systems. Nope, you now have the knowledge to explore complex topics and concepts.

In this section, aptly named the advanced topics section, I am going to teach you about some advanced concepts!

You will be learning the fundamentals of analytics, IoT, and systems integration. By the time you finish this section, you will have the knowledge you need to work with these very important systems.

CHAPTER 13

WHAT CAN ANALYTICS DO FOR YOU?

What's in this chapter

A nalytics is one of the hottest topics on the market right now. Analytics are not only one of the most discussed topics in the smart building market, but they're also one of the most misunderstood topics in the market.

In this chapter, I am going to unpack the topic of analytics. You are going to learn:

- What analytics are and what they aren't
- The different types of analytics
- How trends can and should be used for analytics
- The common platforms for analytics
- The six steps for implementing analytics

I briefly discussed data during Chapter 4 of the IT concepts but being that this is the advanced section I am going to go deeper into the concept of data.

Specifically, I am going to teach you exactly what data is. I am also going to show you how you can adopt a data strategy that will position you to be able to execute analytics whether you choose to do so right now or in the future.

Let's go!

What are analytics?

So, what are analytics?

Analytics are both the results of and the tool used to perform data analysis. There are two main approaches for data analytics. Those are assumptive analytics and non-assumptive analytics.

Assumptive analytics begin the data analysis process with an assumed result. When you look at a dataset from an assumptive perspective, you are often looking at that dataset with a hypothesis. This means that you are stating that you believe the data will tell you something and you are analyzing the data to find out if that "something" is true.

With **Non-assumptive analytics,** you are using several different "patterns" to analyze data and to alert you to any trends. This is different than assumptive analytics because you are not approaching the data with an assumed result.

By the way, the majority of the analytics used in the building automation space today are assumptive analytics.

Regardless of your approach, the process for executing analytics remains the same. Here are the three most common steps that are taken in an analytics approach:

1. You identify the data you would like to analyze
2. You either establish data collection and store that collection as a dataset or you gather data from an existing dataset
3. You analyze information from that dataset

What are datasets?

Datasets are a collection of data, structured or unstructured, that have a common theme. For example, a collection of the VAV box data is a dataset.

Now, what do I mean by structured and unstructured data?

Structured data is data that has a formal structure to it. For example, think about a space temperature measurement. When you read the space temperature from a BAS system, at least here in the US, you get an analog value that is formatted using the Fahrenheit temperature scale.

That's how it is, and it is always that way, it's structured!

Unstructured data, on the other hand, is the opposite of structured data. Unstructured data is data that is not formatted or is unpredictable in its structure. An example of this would be hand-written service tickets. Handwriting is different for each person, and you can't predict where someone will write on the page.

Data breaks out even further into two buckets. These buckets are historical data and real-time data.

Historical data is fairly self-explanatory. You can use historical data to look at events that have happened in the past. Using this data, you can analyze the data to find historical patterns.

Real-time data, often called "streaming data" or "data streams" is data that is being constantly generated. This data requires constant analysis. With real-time data, you are looking for signs of performance against a scenario.

Analytic methods

In the BAS space, there are two main analytic methods that are used to analyze data. These methods are called rules-based analytics and pattern-based analytics.

Rules based analytics

Rules based analytics look at data and alert the person performing the analysis when a specific rule is met. A rule is a threshold or condition that when met will cause the analytics rule to be triggered. This is also known as condition-based analysis.

Pattern based analytics

Pattern-based analysis is a method that identifies patterns in data according to an algorithm. Now don't let the word algorithm scare you, you'd be surprised to know that you use algorithms every day!

Whenever you solve a math problem using a formula you are using an algorithm.

With pattern-based analytics you are going to be looking at the data and seeing if it matches a pattern.

Examples of analytic methods

Some of you may be asking, how is pattern-based analytics any different than rules-based analytics?

Great question.

Rules based analytics are constant. This means that they use hard-coded variables to determine if an event is triggered.

Pattern analytics are fluid. The pattern will adjust based on the inputs being provided to the algorithm. To nail this down I'm going to show you an example of each type of analytics.

Rules based methods

One of the most common rules based scenarios is the deviation scenario. As you may recall from Chapter 2, a BAS system has alarms that can be configured to trigger when a space temperature goes out of range.

Now, what if you had 2,000 VAV boxes and 10% of them were alarming. How could you determine which boxes are the ones you should focus your maintenance efforts on. This is where rules analytics comes into play.

With rules-based analytics you could say "show me the top 10% of the boxes that are ordered by the difference between set point and zone temperature."

You could also sort by how long the box has been away from temperature. Now you have two very valuable data points, comfort and time.

With this information, you can prioritize your maintenance efforts to address the boxes that are furthest from set-point or have been in error for longer than a certain time period.

Pattern based methods

In contrast to the rules-based method, the pattern method is looking for a constantly varying pattern.

A classic example of the pattern based method would be the performance of valves over time.

With pattern-based analytics you could set up a pattern to detect if your valve performance drifts over time which could indicate fouling of the piping or cools.

In this scenario the analytics solution would monitor a slew of points:

- Valve inlet and outlet temperatures
- Water flow (GPM)
- Discharge and mixed air temperatures
- Airflow (CFM)

The analytics solution could read these values to create a baseline of this unit and other similar units. As a system deviates further and further from this baseline, the system will begin to trigger alerts indicating that a particular part of the system is failing.

Trends the original analytics

You may be asking yourself, "Why is Phil calling trends the original analytics?"

That's good question.

The reason behind the title of this section is that folks often forget that they have their own miniature analytics platform built into their building automation systems.

What is this mini-platform you ask?

This platform is the trending and reporting capabilities that most building automation systems come with by default.

As I mentioned in chapter 2, trends give you the capability to log the performance of your BAS points over a given period of time. You can capture trends based on a time interval or change-of-value. Once captured these trends can be sent to a database in batches or streams to be stored for access later.

Now if you know what you're doing, you can implement many of the pattern based analytics methods using your trends. The reason folks utilize analytics is because setting up your trends can be time-consuming and picking out a pattern within your trend reports can require a certain level of expertise.

What kind of patterns can be picked out of trend reports and how can you do that?

As I see it, there are three main patterns that can be found utilizing trend reports. These patterns are:

- Simultaneous heating and cooling
- Improper damper control
- Improper system staging

Simultaneous heating and cooling

Simultaneous heating and cooling is a common pattern that is detected by analytics. What happens is that over time the BAS field controller gets out of calibration. This causes the controller's functions that oversee the heating and cooling to be triggered simultaneously. This results in the system cooling the air just to heat the air back up.

As you can imagine this results in significant energy waste. In addition to this, buildings can spend a significant amount of their cooling or heating load on this issue. This can result in facility teams believing that they do not have enough capacity in their cooling or heating systems.

I've personally been to facilities that were about to purchase a new chiller or boiler to compensate for a perceived lack of capacity.

Once their systems were analyzed, they realized that they had enough capacity. What they perceived as a lack of capacity was a simultaneous heating and cooling fault consuming their capacity.

To set up your trends to detect simultaneous heating and cooling you need to follow these steps:

Step 1: Create trends on the following points:

- Discharge air temperature
- Discharge air temperature set point
- Heating valve output
- Cooling valve output

Ideally, these points will be set to trend every five minutes.

Step 2: Create a trend study or report that allows you to view all of the trends you set up in step one. The timeframe for this trend study can be up to a year if you have filtering capabilities within your building automation system.

If you do not have filtering capabilities within your building automation system, you will want to set up the timeframe for your trend study to be a week or less.

Step 3: If your building automation system allows you to filter. You will want to filter for trend samples where the cooling and heating valve are greater than 0% at the same time. Any systems that fit these criteria are most likely experiencing simultaneous heating and cooling problems.

If you do not have the capability of filtering your trends, you will want to go and manually look at your trend graph. You are looking for times when both your heating and cooling valve are open at the same time.

Improper damper control

One day I was sitting in my work office, and it was hot. It was about 92° outside in Milwaukee Wisconsin, but the inside of the office was 76° with a relative humidity of around 60%. I do not know why it was so warm and humid inside the building but in my experience, improper damper control has led to many hot and humid buildings.

But how do you detect if the damper is not being properly controlled?

I will show you how in these next couple steps:

Step 1: First you need to make sure you're trending the right points (hopefully you all are noticing a trend by now, and yes the play on words was intentional). The points you want to set up depend on the dampers you are trying to analyze.

Typically, you're trying to detect if the dampers for your air handling units are controlling properly. To do this, you will need to set up trends on the following points:

- Outdoor air dampers
- Return air dampers
- Outdoor air temperature
- Return air temperature
- Mixed air temperature
- Economizer enable set point

You will want to trend these points every five minutes.

Step 2: Next you will want to export your trends as an excel file. You will be using this document to perform some basic math.

First, you will multiply the outdoor air damper position and the outdoor air temperature.

OAD * OAT = CALC1

Next, you'll want to take the return air damper position and multiply that with the return air temperature.

RAD * RAT = CALC2

You will then want to add the results of the return air and outdoor air calculations together. That temperature should match your mixed air temperature.

CALC1 + CALC2 = MAT

If your mixed air temperature is much higher than that calculation, then you most likely have outdoor air damper that is leaking through your dampers.

Improper system staging

Improper system staging is a common error within building automation systems. Often called short cycling, **improper system staging** is when a multistage system is either turning on or off too often or not often enough.

You will want to use the following steps to detect this error:

Step 1: Trend your systems using a change of state or change of value trend. You will also set up trends on the following points:

- Controller outputs
- Your sequencer input
- Each stage's on and off timers

Step 2: Now if your building automation system allows for filtering of your trends. You will want to filter your sequencers on and off times based on the on and off timers. This math is a little bit difficult to do, and you will want someone who's very experienced with formulas to help you with this.

However, at the end of this, you will notice that stages are turning on and off when they should not be. You can then go and address either the logical or physical timers that are causing this issue.

Common analytics platforms

There are two types of analytics platforms that I am going to cover. Those platforms are:

- On-premise
- Off premise

The biggest difference between on premise and off premise platforms is where the actual software solution is located. In this section I'm going to teach you:

- What an on premise and off-premise solution is
- The common software capabilities that they provide
- The things you need to be aware of before purchasing one of the solutions

Before I dive into my discussion about on-premise versus off-premise solutions, I want to cover the capabilities that are normally provided with analytics software.

What capabilities are normally provided?

The capabilities provided by on premise and off premise software are pretty much the same. The big difference between on premise and off premise software is that off-premise software can be scaled to handle large analytical loads.

This means that the off-premise hosting environment can be quickly scaled if you add a lot of points to the system. With an on-premise system, you would have to purchase and install more storage or compute.

On-premise solutions

What are they?

In the IT world systems are defined as on premise and off premise. **On-premise** means that the system and its associated parts are located at the customer site.

The benefits to having an on-premise system include:

- Low to no hosting costs
- Direct control of the data
- No outside access required

The cons of an on-premise solution are:

- Higher first cost
- Potential for large data storage requirements
- Management of the software

What do you need to be aware of before purchasing?

When you're looking to purchase an on-premise solution, you need to consider a few things. I've listed out the things you need to consider:

- Do you have the technical capabilities in-house to support the software?
- Do have the storage and compute capacity to run the software
- Can you afford the higher upfront capital cost associated with on-premise software?

Off-premise solutions

What are they?

An off-premise solution is the exact opposite of an on-premise solution. An **off-premise** solution allows the customer to host all or some of the analytics software functionality off-site.

The benefits to having an off premise system include:

- Lower first cost
- Not having to manage the software
- Theoretically unlimited storage capacity

The cons of an on-premise solution are:

- Reoccurring software costs
- Not having direct control of your data
- Relying on another organization's infrastructure and technology

What do you need to be aware of before purchasing?

An off-premise solution is fairly similar to an on-premise solution, but there're a couple of key points you need to consider before you go and purchase an off-premise solution:

- Do you understand and have you accounted for the cost structure of an off-premise solution?
- Does the off-premise provider have a security and data protection process?
- What are the uptime guarantees provided by the off-premise provider?

Implementing analytics

At first, implementing analytics can seem quite difficult. In this section, I will show you how implementing analytics comes down to a set of repeatable, predictable steps.

There are six steps that should be considered before implementing any analytics project.

Step 1: Identify the outcome

It seems obvious that if you want to buy analytics software you should have an outcome that you expect the software to help you achieve in mind. The problem I've seen, though, is that a lot of folks have an unrealistic expectation of what analytics software can do.

It's important to ask yourself, "What you and your stakeholders believe the analytics solution will accomplish?"

Here is how you can effectively identify the outcome(s) that the analytics solution will provide you.

First, outline the problem you believe you have. For example, you have several mechanical systems that keep failing.

Next, detail out how this problem could be fixed. Maybe you expect the analytics software to detect potential failures before they occur?

Finally, describe what success would look like. Perhaps success would look like a 10% reduction in mechanical failures?

The key point is to identify the problem, describe the problem, and then agree upon what a successful outcome would look like.

Work with the analytics provider to ensure that the analytics software can provide this outcome.

Step 2: What do you need to make that outcome happen?

The next question is, "What do you need to make this outcome happen?"

In the previous step, I gave the example of reducing the mechanical failure rate by 10%.

Imagine that these mechanical failures were being caused by fans and pumps that were being run for too long.

Taking this a step further, let's say you have a way to measure the performance of the fans and the pumps. You have identified a pattern that says if the pumps run for a certain amount of time and during this period, the amount of water they can move decreases, the possibility of an eminent failure exists.

What do you need to measure the scenario?

You would need to be able to monitor the output and runtime of the fans and pumps. You would also have to determine, working alongside the analytics provider, how much of this data you would require and how frequently you need to sample the data.

Which brings us to step three.

Step 3: Do you have what you need?

This step is quite simple if step two was done properly. In this step, you are simply verifying that you have the points you need to measure the runtime and output of the pumps and fans.

Pretty cut and dry right?

This is the point at which many analytics proposals die.

What happens is folks realize that they have to add a whole bunch of new points and capabilities to implement analytics. Because of this, the cost of the analytics solution goes through the roof, and the project never happens.

However, isn't this better than the alternative?

Imagine if you didn't verify that you have the data and points that are needed to implement an analytics solution. What would the cost be in that scenario?

When I tell you this book, will easily save you at least its cost. This is how.

Performing this step and avoiding this scenario will save you thousands.

If you get past this point, and you still want to go forward with the analytics project, you will want to move to step four.

Step 4: How can you get what you need?

So you have the list of points that you created in step three. Now the question becomes, "How can you get the data?"

In step four you need to answer several questions.

Will you be grabbing information from the database?

Do you have access to the database?

Are your points being trended and stored in the database right now?

How much data storage will you need and how often do you have to sample the data?

Can your system handle the amount of traffic generated by all of the data samples you have to capture?

These are important questions that you will need to ask yourself if you are moving forward with an analytics project. These questions when not properly addressed can introduce significant risk and cost to the project.

However, there's an even more important question that needs to be asked, and I will cover that in step five.

Step 5: Who interpret the results?
You've got all this data, awesome!

Who's going to make heads or tails out of it?

No seriously, so often folks will focus so much on the technical aspects of analytics that they don't even think about who's going to interpret the results. Analytics can point out several key pieces of information, but at the end of the day, a person still has to make a decision on what actions to take based on the results.

The question is, "Does this person have the skill set to interpret the analytics?"

If the answer is no, then how can you accomplish this?

You can accomplish this in one of two ways.

The first way you can do this is to hire someone who has the skills to interpret this data. The second way you can do this is to pay the analytics provider to interpret the data for you and provide you recommendations on what actions to take.

Now, you may be saying to yourself "But Phil isn't that what the analytics software is supposed to do?"

Yes, and no.

Analytics software will identify patterns and make recommendations, but it's up to you to interpret these recommendations based on the potential impact they may have.

Step 6: How will you implement the results?

Let's say your analytics software suggests that you need to go and replace the valves on 10 of your air handling units.

How would you do this?

No seriously, if your techs are constantly reacting to troubleshooting calls, who is going to implement the changes that the analytics system suggests?

This is the main challenge I've seen on several of the analytics projects that I've implemented.

Everyone was so excited, "We're getting analytics yay!"

However, no one stopped to think about who was going to implement the actions required to generate the results. Folks got upset when they realized that analytics were not an easy button.

Many times these people felt tricked into buying an analytics solution. That's why it is so critically important to have this conversation up front. Take the time and define who will implement the actions recommended by the analytics solution.

If you remember anything from this chapter, remember that analytics are only as good as the person taking action on their recommendations.

Chapter 13 quick summary

Nice!

You now have a much greater understanding of analytics and how they can impact your business!

In this chapter, I took you deep into the world of analytics. I showed you what analytics are and how they can work for you! I even went and walked you through what using analytics would look like in real-life.

Next, I showed you how you could use your trends to perform analytics. I taught you how you could use your trends to identify the three most common "faults."

I then explained the difference between off-premise and on-premise analytics platforms and the pros and cons between them. I closed this chapter out by giving you my 6 step process for implementing analytics in your business!

In the next chapter, I am going to unpack the most misunderstood topic in the BAS space, the Internet of Things...

CHAPTER 14

IS IOT FOR REAL?

What's in this chapter?

The Internet of Things (IoT), is gaining massive popularity in the world right now. IoT is one of those things that at the surface level makes great business sense. I mean who doesn't want to go and capture more data to drive effective, efficient action within their business?

But as with any new technology IoT isn't without its challenges. IoT is a topic that several of my customers routinely ask me about. Often I get questions like:

How do you apply IoT to a building?

How do you design IoT?

Is IoT real?

It is these exact questions that this chapter will answer. By the time you are finished with this chapter you will know:

- What IoT is
- The current state of IoT in the market
  - o Key IoT terms you may be unfamiliar with
  - o IoT landscape
  - o IoT functional areas
- IoT design patterns

Are you ready to learn about IoT?

I sure hope so. Let's begin!

What is IoT?

According to the Internet, and we all know the Internet never lies, the term IoT was coined by Kevin Ashton in 1999 when he was working for Proctor and Gamble.

The term IoT stands for the Internet of Things.

Now if you're like me you may be asking yourself, how is this any different than building automation systems?

After all, building automation systems consist of a network of sensors and control devices that communicate with one another to drive the control of building systems.

I've asked this same question and here are some of the answers I was given:

- IoT will provide you with more data from your sensors
- IoT will allow devices to communicate with one another
- IoT will allow your "things" to make decisions without you being in the middle

Those answers are well and good, but the reality is a properly designed BAS already does these things.

Let's be real for a second...

I went and interviewed several facility managers during the process of writing this book.

The feedback was that they already have too much data.

Well, that's ok, because IoT will enable these devices to talk to one another and that will reduce the amount of data the user has to process right?

Maybe, then again, probably not.

Here's why...

BAS controllers already communicate with one another. They've been doing that for years.

Ok then, surely the fact that IoT will allow your "things" to make decisions without you being in the middle is a key selling point of IoT, right?

You may be noticing a theme here.

Guess what, devices that make decisions without you having to command them already exist.

Don't believe me, consider discharge air temperature resets based on the average of all your space temperatures. Sure looks like inter-device communication to me!

Ok, so then what is the benefit of IoT?

The benefit of IoT, as I see, it is that IoT provides the ability to process massive amounts of data closer to the edge of the building (more on the edge later)

In this next section, I am going to look into why the ability to process massive amounts of data closer to the edge of the building is valuable.

The ability to process massive amounts of data closer to the edge of the building

You may be shocked that I didn't call out analytics as a key feature that makes IoT valuable. The reason I didn't call this out is that analytics can and has been done without IoT.

So then what does this statement about processing data closer to the edge of the building even mean?

I think the best way to answer this question is to break this statement down piece by piece.

If you recall in chapter 13 I mentioned how data could be collected from a building automation system in batches or real-time streams. No matter what form of data collection is done, the data still needs to go back to a central application where it can be processed.

It's only when the data is processed that you know if the data is valuable and what information the data provides to you.

Because of this in some analytic applications, there can be delays of up to four hours. This delay is caused by the time it takes for the data to flow up to the cloud, run through the analytics and be communicated back down to the user.

The big promise of IoT is that all of the data that is being captured will be able to be processed at an edge gateway.

Essentially the data will flow from the sensor to the edge gateway. The data will then be processed at the edge gateway. Once this processing is done, the data can be discarded, communicated to the user, or sent to the analytics application for further analysis.

I realize it's hard to visualize what I just said so imagine this scenario.

You have a fan coil unit with a heating valve and cooling valve. For some reason, this unit is commanding the heating and cooling to open at the same time. Now normally you would have to wait for this data to be fed to your analytics software or for someone to analyze your trends to realize simultaneous heating and cooling is occurring.

However, with IoT, the data from the valves can be processed at the edge gateway that resides near the controller. This analysis can then immediately identify simultaneous heating and cooling. In the future, IoT solutions will even be able to issue an appropriate command to the controller.

If you stop and think about all the different scenarios like this that could happen. You can begin to see how the Internet of Things could significantly decrease the amount of reactive maintenance facility teams have to do.

The ability to communicate machine to machine

In my humble opinion machine to machine, communication has been going on for the past 15 years. Building automation field controllers today can communicate between one another using network variables.

So, how is this can be any different with IoT?

Here are my thoughts on this.

Right now if you want to add a lighting system or specific sensor to a building automation system you have to go and discover the system, address the points, name the points, and assign them to the right systems.

An IoT scenario that I've heard promised but have yet to see delivered is the automatic discovery, addressing, naming, and allocation of sensors and systems.

In a truly hands off IoT scenario, a control system would simply need to be connected to the network and it would auto commission itself.

Several things would have to happen within the industry for this to take place.

First, you would have to have a way to determine the location of the device and the relationship to other devices. Next, you would have to have a way of understanding the sequence of operations and applying that to the IoT device.

Finally, you would have to have a way to add the device into the building automation system.

I'll tell you. I work with some of the most advanced building automation technology at some of the most complex building automation projects, and despite what the marketing efforts of some companies will tell you, the BAS world is not close to doing this.

IoT will allow your "things" to make decisions without you being in the middle

This is the aspect of IoT I'm most excited about, and here's why.

As you may or may not know building automation monthly's mission is:

To provide the best online Building Automation training to prepare 20,000 people to enter the Building Automation space by 2025

The reality is the amount of devices and sensors being added to buildings is massively outpacing the capacity of building automation professionals. If we are to have truly smart buildings, then the buildings themselves will need to perform a lot of the day-to-day tasks without involving a building automation professional.

However, if you've spent any amount of time running a building automation system or performing building automation system installations or service, you know that there is nothing "standard" about a building automation system.

And therein lies the rub, as long as building automation systems are designed to implement sequences that are unique to each and every building, we will run into issues with self-executing control systems.

This will most likely be the last capability that IoT can address.

I know there's been a lot of buzz around building automation systems that will supposedly analyze a building automation system for faults and automatically correct these faults.

Where is this being done?

The problem with automatically correcting faults is that your fault profiles could be completely wrong. If you go sit down and have a beer with the technical lead of any analytics solution, they will tell you that while their analytical solutions can detect most faults, they would not recommend that the system automatically corrects the faults.

The reason for this is the variety of control sequences that exist.

I can hear some of you saying "Well, Phil surely some faults can be fixed automatically, for example, an override could be automatically fixed."

But could it?

Could you automatically release overrides?

What if a device was overridden because someone was working on it?

Would you want to be the person responsible for a system failing or someone getting hurt because you released the override?

The Current state of IoT

The IoT market, no matter what anyone tells you, is fragmented.

There are so many competing protocols and frameworks that the market is trying to get vendors to adopt, that it can be impossible to keep up. In the next couple sections, I will discuss the IoT landscape and the IoT functional areas.

But before I do that it would probably help if I level set on some of the common terms you'll hear folks use when discussing the Internet of things.

Common IoT terms

Just like any other technology the Internet of Things has its own terminology or" lingo," that folks use to describe it.

In this section, I'm going to help you understand some of the most common terms that "IoT experts" use (if you're wondering why I put the term IoT experts in quotes, it's because this market is still so immature I'm not sure how anyone can consider themselves an expert).

Now, I want to be clear. This is by no means an exhaustive list of every term that you are going to hear when you are talking to people about IoT.

However, having sat in many, many, meetings about the Internet of Things, I can tell you these are the most common terms you will hear.

Actuator- when you hear the term actuator used about the Internet of Things this is not about a valve or damper actuator. Rather, an **actuator** in the Internet of Things is any device that performs an action.

Edge-Device- with the Internet things comes the concept of the edge device. The edge device exists on the edge of the network. **Edge devices** typically serve to collect data from sensors and provide commands to actuators.

Ecosystem- with the introduction of the Internet of Things comes the concept of an ecosystem. Previously reserved for software vendors, the open nature of the Internet of Things makes it attractive for hardware developers to provide pieces and parts.

Framework- a framework is a blueprint for creating an Internet of Things architecture. There are three primary frameworks that are used within the IoT market.

Gateway- all of the Internet of Things actuators and sensors tie into devices called gateways. These gateways exist across multiple zones. The purpose of a gateway is to capture, filter, and forward data from sensors and actuators to the Internet of Things applications.

M2M- M2M stands for machine-to-machine communication, and is one of the core concepts of the Internet of Things.

Sensor- in the Internet of Things world, sensors are any devices that provide feedback to the Internet of Things network.

Zone- a zone is a specific layer of functionality used to segment certain parts of the Internet of Things. It is common for an Internet of Things architecture to have several zones based on the level of functionality that is desired.

For example, you may have a sensor/actuator zone that is specifically focused on data collection and device control.

The IoT landscape

To say that the IoT landscape is fragmented would be an understatement. Every day it seems that a new IoT company is popping up. However, in the BAS space, there is nothing new and innovative that is being created.

For example, several new IoT companies have popped up claiming to be the "next generation" building automation system.

However, when you dig into it, you realize all these companies are doing is consolidating data and providing a pretty looking graphical user interface with some basic analytics on top of it.

The reality is IoT still requires the operator to run the building, fix problems, and analyze data.

The promise that IoT would eliminate all these tasks has not happened.

Now don't get me wrong, I believe that the building automation industry is ripe for innovation and change. However, don't buy into all the hype that is being spun out in the market.

With that being said what does the IoT landscape look like today?

Rather than diving deep into individual companies I am going to focus on functional areas.

The reason for this is most of the companies that I would mention probably will not be here a year or two from now.

IoT functional areas

The IoT market breaks out into three functional areas. These areas are:

1. Edge Analytics
2. Protocol gateways
3. Visualization software

Edge Analytics

Edge analytics is just as it sounds. Up until recently, information had to be collected using centralized analytics. This required data to be collected, transmitted, stored, and processed.

Because of this process, the building automation system operator could have a delay of up to a day depending on how the centralized analytical solution worked.

With **edge analytics**, data will be processed closer to the actual sensor. The analytics software will process the data and then decide whether the data needs to be sent up to a centralized analytics solution for further processing,

This will reduce the amount of time to analyze data and will reduce the amount of bandwidth and storage required for data.

Protocol gateways

IoT has brought about the introduction of new software and hardware providers. These new software and hardware providers are bringing with them new protocols and APIs.

The amount of different communication formats is only going to increase. Because of this, there needs to be a way to process all of the data and normalize the data before it arriving at the graphical user interface.

Enter, the protocol gateway!

The **protocol gateway**, often called a messaging bus, will connect to all of the different systems and will "normalize" the data that is being collected into a specific communication format.

In addition to normalizing communication, the protocol gateway will also normalize the data model.

The biggest problem IoT is going to face has to do with conflicting data models. For example, lighting providers like to use a zone model. With a zone model, the lighting can be assigned to multiple different spaces.

In the building automation world, the space model is often used. This means that a specific controller or sensor is connected to a specific space.

If you try to match up the lights (using a zone model) to a building automation system (using a space model), you will have what is called a data model conflict.

This will result in, data that "doesn't match up." Currently, data model conflicts have to be resolved manually through programming using something called a data adapter.

If IoT is to scale, these protocol gateways will need to be used.

Visualization software

The final functional area of IoT is visualization software. Do not confuse visualization software with a building automation system graphical user interface.

Visualization software exists to help analyze, process, visualize, and report on all of the data that is being collected. I've worked on some rather complex projects.

One of the consistent themes that I have heard from customers is that there is an increasing sense of overwhelm due to the large amount of systems that are being integrated. This is because there is too much data available.

For the building owner to make quick and accurate decisions, this data needs to be formatted in a way that is easily understood.

This is the area in which IoT solutions are starting to mature. There are several cloud-based visualization software providers.

IoT design patterns

Forgive me, but I'm going to go a little deep in this section. I know, I know, this is a primer book, just humor me and read this section.

In this section, I'm going to help IoT become real for you.

I'm going to do this by talking through a concept called the design pattern. A **design pattern** is a representation of a design that you can use to build a system architecture.

Think of design patterns as the mechanical plans for IoT. As I see it, there are four design patterns that apply to the building automation space. These designs are:

- Edge data collection
- Publish/subscribe
- Dumb IO with cloud hosted software
- Dumb IO with edge software

Edge data collection

The edge data collection design pattern involves IoT Devices that are located on the "edge" of the building automation system. These edge devices will feed their data up to a gateway that collects the data and forwards it to a centralized location.

Edge is a term used to describe where devices are located.

Publish/subscribe

A publish/subscribe pattern, also known as the pub/sub pattern, is a design pattern where the IoT devices publish data that other IoT devices can subscribe to.

Dumb IO with cloud hosted software

With this design pattern, a standard controller with input and output capabilities is used. No logic resides in the controller, which is where the term "dumb" comes from.

The software that commands the outputs and reads the inputs resides on a server in the cloud.

Dumb IO with edge software

This pattern is almost the same as the cloud hosted pattern except that the software resides on a gateway device close to the edge. This allows for logic that is more complex or logic that requires a lot of communication with the I/O devices.

Chapter 14 quick summary

This was a short chapter. Because…

IoT is still largely vaporware. There it is, I said it! I know, I know, for you sales folks out there I'm killing your #1 message right now.

Here's the deal, I made way more sales in my career calling out the truth than I ever did promoting hype. I mean seriously, when you visit the websites of several controls manufacturers you will see how they are positioning their BAS as an IoT platform.

Come on!

Now don't get me wrong, I know folks in marketing and communications, they are awesome folks! They are not trying to mislead anyone. The thing is, I know, and now that you read this book you know as well. A lot of the "IoT" features are nothing new.

So, you slapped a Wi-Fi antenna on your controller, or you're using mesh to pick up sensors great!

You're doing analytics in the cloud, awesome!

The thing is, none of those features are "IoT." They're just enhancements on an existing product line.

Ok, ok…

I'm getting off my soapbox.

So what did I cover in this chapter?

In this chapter, I taught you what IoT is and what it isn't. I took you through the current state of the market, and I showed you the three key ways I believe IoT will impact the BAS space.

I also broke down a lot of the "lingo" that you hear folks using in the IoT space. No more sitting in meetings while folks speak a language you don't understand!

I then described the IoT landscape and the design patterns that you can use to implement IoT on your projects.

Ok, folks, it's the moment you've been waiting for. Everything I've taught you up to this point has led to the next chapter!

CHAPTER 15

INTEGRATION, EVERYONE'S FAVORITE WORD

What's in this chapter?

I was thinking of how I was going to write this chapter and I had so many different thoughts in my head. At first, I thought, you know folks just need to know what systems integration is and the important things to look out for.

Then I got to thinking," Well maybe what folks need to know is how to execute a systems integration project"?

So after mulling this over for a ridiculously long amount of time, I figured why don't I talk about both?

And here you are.

Welcome to the most anticipated chapter of the book, systems integration.

So what am I going to teach you in this chapter?

Good question. Here's what you're going to learn:

- What is systems integration?
- How can you define the scope of an integration project?
- How can you ensure an integration project gets executed properly?
- How can you support an integrated solution after it's installed?

Over the next several pages, I am going to take you on a journey through the topic of systems integration.

By the time you're done with this chapter, you will understand what systems integration is and you will know how to scope, manage, and support integrated systems.

Alrighty, let's dive in...

What is systems integration?

Systems Integration is the act of taking two or more separate systems and integrating them so that they act as a single solution.

The problem is most people confuse systems interfacing with systems integration.

Which brings me to the next section.

Systems integration vs. system interfacing

Systems interfacing is the act of combining two or more systems using a common format. When most folks say they are integrating systems, they usually are performing systems interfacing. An example of this would be when BACnet/IP rooftop units are interfaced with a building automation system.

This is not integration, rather, this is interfacing the rooftops to your existing system.

So then what would be an example of systems integration?

An example of systems integration would be taking the trend data from multiple building automation systems and combining them into a single master database.

You see, when you're performing **systems integration,** the result is a different system than the systems that were integrated together. This might seem confusing right now, but I promise you it will begin to make sense as you read through this chapter.

The three levels of systems integration

Before I get started with the how I need to define the what. And that what is best defined by the three levels of integration.

Often when I sit around the table with customers they see integration simply as the act of connecting two systems together.

However, there is much more to integration than just connecting systems. As I see it, there are three levels of integration.

This is a critical concept and if you pick up nothing more from this chapter than this section I will consider it a success.

I have seen a lot of folks try to approach systems integration without understanding the level of integration they are trying to perform. As you will see, these different levels require different approaches to the systems integration process.

These levels also require different levels of expertise, materials, and software.

These levels are:

1. Process integration
2. System integration
3. Software integration

Process integration

Process integration is something you do every day, and you probably don't think about it. Whenever you take two tasks and combine them to produce a single task that is different than the original two, you're performing **process integration**.

An example of this would be combining two separate processes together to produce a greater "master process."

For example, let's say that you have 10 VAV boxes supplied by an air handler. The sequence of operations resets the discharge air temperature for the air handler based on the zone temperature of the VAV's.

You also have a central plant that has a primary and secondary chilled water loop. This central plant provides a constant 42°.

Both of those examples are individual control processes.

In process integration, you combine these two processes to form a new process. An example of this would be taking the valve position of all of the air handlers and resetting the chilled water supply temperature setpoint.

System Integration

Systems integration, for which this chapter is named, is the process of integrating two separate systems to create a new system.

For example, if you were going to take two building automation systems and combine these building automation systems together to form a single building automation system. You would be performing systems integration.

The reason this is systems integration and not interfacing is because the two building automation systems are becoming a separate new building automation system. In systems interfacing, you would simply tie one of the building automation systems into the other.

Software Integration

Software integration is where you combine software applications or the functionality of software applications to create a new application or new set of functionality.

An example of this would be taking data or functions from multiple application programming interfaces and combining this data and functionality together to create a new application.

Hopefully, you see that when you combine two or more things together to create a new thing you're performing systems integration. But, when you're simply pulling one thing into another you're performing systems interfacing.

Why should I care about systems integration vs. interfacing?

You may be wondering why you should even care about systems integration versus interfacing.

You may be asking, is all of this simply semantics?

Am I simply standing on a soapbox preaching to you?

No, not at all.

The reason this is so important is that when you perform systems interfacing, you are not creating a new thing, you're simply pulling an existing thing into another existing thing. This means that if everything goes wrong, you can still get people who understand how both "things" work to troubleshoot the "thing."

However, with systems integration, you're creating a new system.

You are the expert because you created this new system. This means there's no one to call on, it's just you.

Now, this isn't a reason to avoid doing systems integration. Systems integration can provide great benefits to building automation professionals.

This is, however, a reason for you to have a process to utilize when you approach systems integration.

How can you define the scope of an integration project?

The first thing you need to do whenever you are approaching a systems integration project is to level set on the expectations and define the scope.

How can you do this?

To define the scope of an integration project you first need to identify the outcome(s) that your stakeholders want to achieve.

As I mentioned earlier in the book, stakeholders are folks(s) who have a personal stake in the success of the project. This could be a facility manager, a building owner, contractor, or even a system integrator.

There are multiple methods you can use to discover the outcomes that your stakeholders want to achieve, most of which I have already described.

In my opinion, the best way to discover the outcomes your stakeholders want to achieve is to have a meeting where you identify the key metrics or processes that you want to improve. Once you've done this, you simply need to detail out how you want to improve them.

During this meeting, you will want to have a subject matter expert for your technology systems involved. The subject matter expert can advise the group around what technologies can be used to achieve the outcome the group desires.

If you find that technology can indeed be used to achieve the stakeholder's outcomes, you will want to move onto the next step. It is during this step that will begin to evaluate the financial and operational costs of creating systems integration project.

As you can imagine, there's quite a lot to that evaluation. The good news is that I have already covered this process in the "Upgrading your BAS chapter."

Once you have agreed upon the outcome, designed a project, and evaluated the project for an acceptable return on investment, you can begin to look at how you will implement this project.

Using the definition of system integration that I provided earlier in this chapter, you can evaluate if you will need to integrate systems as part of this project.

If at any point you are unsure of the need for systems integration, you should look at engaging a systems integration subject matter expert.

Performing systems integration

Once you've adequately defined the scope of your project, it's time to move onto the actual project itself. In this section, I'm going to lay out a step-by-step process for you to follow to execute your systems integration projects.

After doing hundreds of systems integration projects, I've identified eight critical steps for performing systems integration projects. The steps are:

- Step One: Creating a systems integration use case
- Step Two: Identifying and diagramming the existing systems
- Step Three: Using a gap analysis to detail out the new systems
- Step Four: Determining the methods to close your gaps
- Step Five: Detailing out the physical and logical integration points
- Step Six: Creating an integration map
- Step Seven: Detailing out the data model and system requirements
- Step Eight: Developing a sequence of operations

Step One: Creating a systems integration use case

In step one you will take the scope of work that you created as well as the outcome(s) that you agreed on with your executive sponsor and used that information to create a use case.

A **use case** details out how the system(s) are going to achieve the outcome you decided upon. The use case begins by detailing out an "Actor."

An **actor** is a person or system that performs a task.

The "**modeler**" is the person who creates the use case. The modeler will detail out how the "actor" interfaces with the systems and performs the tasks related to the use case.

There are multiple different ways to lay out a use case, and I am not going to go into the different methodologies around use case creation in this book.

However, I will tell you that the use case needs to define the following information to be effective:

- The relationships between the user(s) and the system(s)
- The actions by both the user and the systems that will be taken to achieve the outcome
- The direction in which the actions flow

Step Two: Identifying and diagramming the existing systems

Now that you have built out your use case, you can use it to identify the existing systems that are required to perform the use case. Now, this may seem a bit counter-intuitive.

After all, how can you build the use case if you haven't identified the systems yet?

I want to make sure I'm clear on this one. In the previous step, you identified the systems for the use case. In this step, you are identifying the existing systems that you have installed right now.

It is at this time that you're going to want to go back and look at the systems you have in your current environment. As you identify the systems that you currently have in place, you will want to collect this information and store it. The systems that you already have installed are called your **as-is architecture**.

Now that you have a list of your existing systems you will want to identify the specific systems that you will need to perform the use case. This will be called your to-be architecture. A **to-be architecture** is what you want your architecture to look like after the system integration project.

The purpose of systems integration is to take your as-is architecture and bring it to the level of your to-be architecture utilizing as few new systems as possible.

Step Three: Using a gap analysis to detail out the new systems

At this point, you have created a use case, identified your as-is and to-be systems, and now it's time to identify the gaps.

What is a gap you ask?

A **gap** is a difference between the systems or capabilities that your as-is architecture has and the systems or capabilities that your to-be architecture needs. That's a mouthful so let's unpack that.

As you recall your as-is architecture, is made up of the systems that you currently have installed in your business. These systems have specific capabilities. Your to-be architecture requires a specific set of capabilities as well.

If the capabilities that your as-is architecture provides do not match up to the capabilities required by your to- be architecture, then you have a gap.

The purpose of this step is simply to identify the gap(s), at this point in the process you're not trying to come up with a solution for the gap(s).

Step Four: Determining the methods to close your gaps

Figuring out the best method to close your gaps is really what systems integration is all about. As I said, earlier systems integration is the act of bridging the gap between your current capabilities and the capabilities you want.

Here is how you do this.

First, you get a sheet of paper.

You want to lay out your as-is systems and capabilities on the left side of the paper. Next, you want to take your to-be systems and capabilities and lay them out on the right side of the same sheet of paper.

Then you will draw a line from each of the as-is systems or capabilities to the corresponding to-be system or capability. If an as-is system or capability does not exist that is a gap that will need to be addressed?

You can address this gap by determining whether you need to implement a software, hardware or operational solution.

- **Software solutions** are things like middleware and integration gateways.
- **Hardware solutions** are things like controllers and hardwired connections.
- **Operational solutions** are things like hiring personnel or contractors who have experience with your capability gap.

In the next step, you will detail out the physical and logical integration points that will connect the different systems and capabilities.

Step Five: Detailing out the physical and logical integration points

Moving right along. So far I've covered use cases, system diagramming and gap analysis.

Now, what are you supposed to do with all this information?

Well… now that you have analyzed your gaps and figured out how you are going to close those gaps it's time to connect the dots.

This is the most important step you will take during systems integration.

Here's the deal, a lot of folks will get to this point, and they'll feel that they don't need to detail out the physical and logical integration points.

After all, if you've done a proper gap analysis why would you even have to plan out your physical and logical connections?

It's that exact thinking that gets so many folks in trouble!

I like to call this intellectual laziness.

You see, folks think that since they were smart enough to get to this point, everything else is going to unfold perfectly. Unfortunately, that couldn't be any further from the case.

You can have the perfect gap analysis that details out every gap and its solution. But if you don't show how to connect those solutions together, then all you've got is a fancy design with a bunch of holes.

Think about it this way, if I gave you a bunch of Legos and told you to build a brick house, you would start by connecting the Legos together to build a house.

Right?

Now, what if I put all of those Legos in a bag and told you to shake the bag until the house was put together. You could shake all day long, but I'm reasonably certain that a house would not be created.

That's how it is with integration points. You can put the right systems into your integration project but if you don't detail out how the systems will be connected nothing will happen.

Well, my friends, that BS ends right here. In these next couple paragraphs, I'm going to teach you my step-by-step process on how to map out your physical and logical integrations.

Step 1: Identify your physical connections

People tend to discount the importance of identifying the physical connections that need to exist on a systems integration project. I'm not sure why this is, but it happens so often that it's an area that I need to address.

If you do not identify the physical connections between your individual systems, you will be left at the mercy of whoever installs the systems. This could have really big ramifications once your entire system is installed.

For example, if a critical system is installed on a network that has a lot of video traffic, you could potentially have messages for your building automation system that do not get delivered due to a large amount of traffic.

I've seen this happen time and time again where control systems are put in a highly congested network segment because the designer did not take the time to plan the layout of the physical connections.

Step 2: Identify your communication flow

Once you have your physical connections identified you need to understand how your systems data will flow across these physical connections.

For example, if you are going to be consolidating all of your trend data onto a single device and then sending a large data batch from that device to another device, you will need to account for that traffic.

Often the data flow aspects of control systems is missed.

This results in control systems that share network or other information technology resources. I recall a project with a major data center provider that for three years was plagued by unreliable connections with the other facilities that this company managed.

When I was finally brought in, one of the things I noticed was that communication failures would occur at the same time every day.

After doing some digging, I realized that the WAN link that was being used to share information from the data center to the offsite buildings was also being used to transfer database backups to an off-site storage facility.

Because the communication flow was not properly identified and the appropriate physical connections were not allocated this customer suffered from communication reliability issues for three years.

Step 3: Map your logical connections

However, it's not all about the physical connections and data flow. You can go and perfectly plan out your physical connections and data flow and still have issues. This is because there's a software portion to almost all systems integration projects.

What do I mean by software portion?

If you recall, earlier in the book, I discussed the concept of protocols. One of the things I briefly discussed was that protocols have priority levels. To make sure that systems are not constantly overriding one another, the developers of these protocols created a way to prioritize certain messages.

When you're performing a systems integration project, one of the areas you need to consider is what system will have priority over the other systems?

Often this is ignored, and the resulting outcome is that the project team is forced to remain on the job troubleshooting issues long after the project is complete.

Step Six: Creating an integration map

Even though you have mapped out the physical and logical connections that need to be made, it may still be difficult for the average person to visualize what you have done.

To effectively communicate with all of the people who are involved in a system integration project, you need to put your designs in a format that the project team can understand.

The best format I have found for this is an integration map.

Now there are multiple different ways to create integration maps, and the way I will teach you is my preference. With that being said, just because this is the way I prefer does not mean it is the only way. So by all means, if you have a different way of mapping out your integrations that works for you and your project team, then I encourage you to use that method.

When I am creating an integration map, I like to start by listing out the physical and logical connections that I identified in step five.

In the next section, I am going to show you the physical and logical connections I would identify for a simple integration between a lighting and building automation system.

Physical connections:

- Building automation system to network switch (Via CAT5E)
- Network switch to lighting system (Via CAT5E)

Logical connections:

- Building automation system to lighting system (BACnet/IP Priority 12)
- Lighting system to building automation system (BACnet/IP Priority 14)
- Lighting system to lighting devices (proprietary protocols)

These bullets are meant to be read from left to right and show the physical and logical connections associated with the integrations.

Now, I need to take these connections and turn them into a graphical representation that the average user can understand. For this, I tend to utilize Microsoft Visio Professional.

Microsoft Visio has stencils dedicated to both physical and logical layouts.

Physical diagrams

For the physical layout, I prefer using the ***Detailed Network Stencil***.

Since I've laid out the physical connections already, I simply need to represent the connections in my Visio drawing. The result should look something like the image below.

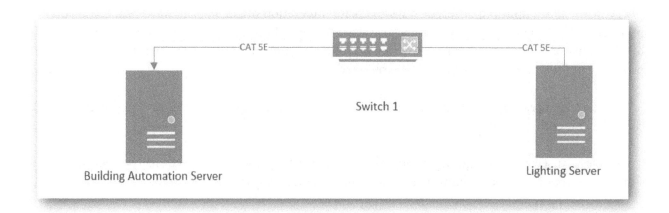

Even though this is a relatively simple drawing, you can see exactly how the building automation server and lighting server are connected to both the switch and to each other. System integration projects can get rather complex.

By laying out the actual connections physically on paper, you now have a common frame of reference from which to communicate.

Logical diagrams

Next, you have the logical diagrams. You will follow the same process in creating the logical diagram mapping.

The stencil I like to use for logical diagrams is the ***Data Flow Diagram***.

Now I will go and create a drawing using the logical connections I laid out earlier to represent these logical connections.

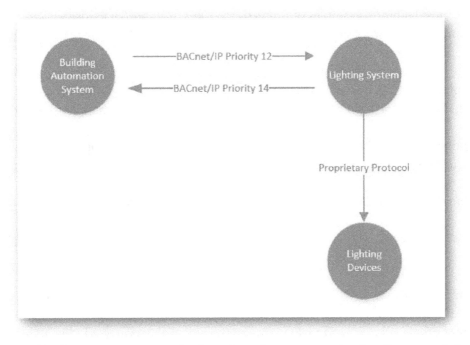

As you can see, this diagram represents the flow direction, communication format and the systems that are being communicated to and from.

I want you to realize that while these examples were quite simple, systems integration can get rather complex. Because of this, it is even more critical that you properly map out your physical and logical connections.

Step Seven: Detailing out the data model and system requirements

Usually, you will be performing this step with the help of a database architect. Now that you have the physical and logical connections mapped out is important to identify the data that will be shared between the systems.

It is also important that you identify the system requirements at this point in your integration project.

Identify the data that will be shared

I've seen countless times where folks purchased "integrated systems" only to realize that the data points these systems shared among one another were not what the person who purchased the system needed.

One of the most common examples of this is purchasing, BACnet/IP cards for equipment. In the past, equipment manufacturers were notorious for creating integrations that barely exposed enough points to make the system functional.

For example, a rooftop manufacturer who will remain nameless would commonly expose only the system enable, discharge or zone temperature and the discharge or zone temperature set point.

Their argument was that providing anything more would make it too complex for the user. As you would imagine many users were upset with this but once the equipment is on the roof, there's not much you can do.

That's why it's important to verify that you can get the data you require.

Here's how you can ensure that you have a list of the data being shared.

Step 1: List out the points you need

Step one is quite simple, all you need to do is list out the points that you would like to see from the system you're integrating. You will want to do this for every physical system you have.

Step 2: Have the manufacturer verify that they can provide these points

Now you will need to contact each of the manufacturers of the systems you have mapped out to verify that they can provide the points you have listed, using the protocols that you have listed out on your logical diagrams.

Step 3: Verify that the other systems can consume those points

Finally, you will want to work with the other system manufacturers to ensure that their systems can indeed map-in, also known in the IT world as consuming, the data points that are being provided.

List out your system requirements

At this point, you're so close to being done. It may be tempting just to go purchase all the systems and start executing the project.

But, here is my "fatherly advice," slow down. This next tip I'm about to teach you will save you so much time, that it will pay for the cost of this book ten times over.

When you are requesting, the information that I listed in the previous section titled "Identifying the Data that will be shared," you will want to ask for the system specifications.

When I say, system specifications, what exactly are you looking for?

You are looking for the following information (while this is not an exhaustive list, it will get you 90% of the way):

- Servers
  - Operating system
  - Memory requirements
  - CPU requirements
  - Required software and conflicting software
  - Storage requirements
  - Virtual or non-virtual server
  - Database requirements
  - Architecture (client/server, micro service, etc.)

- Network requirements
  - Bandwidth
  - IP addressing
  - Routing

I don't expect you to be an expert on each of the bullets I just listed out, although you should be familiar with these terms from chapter 4…

Make sure you ask these questions and request this information. Doing this will help you to shortcut the information gathering process.

This next step is going to surprise you.

Step Eight: Developing a sequence of operations

You may be wondering, why I waited until the second to last step to include developing a sequence of operations?

The seven steps you've taken so far were focused on refining the use cases that you initially started with. As you went through each of the steps, you identified systems that simply could not be integrated with one another.

Because of this, you may have to change your sequence of operations. This is why I recommend creating your sequence of operations last.

Here is a high-level overview of how to approach the challenge of creating a sequence of operations.

You should begin by adjusting your use case based on the information you have found out so far.

Next, you will want to read your use case from left to right. As you follow the main success scenario (the way the use case should work), you will want to write down each step in an electronic document.

Once this process is done, you will have the foundation of your sequence of operations. Next, you will want to tweak this sequence of operations to define any set points or priority levels.

At the end of this process, you should have a solid sequence of operations.

In the next section, I am going to detail out the how to make sure your systems integration project gets executed properly.

How can you ensure an integration project gets executed properly?

As you may recall, in chapter 13 I highlighted how there is a risk of failure any time you make modifications to an existing system. There are also risks that you will encounter as you execute your systems integration projects.

In this section I'm going to list out the four risks that you need to look out for and how to avoid or greatly reduce the chance of these risks occurring.

The four risks you should look out for are:

1. Failing to put systems in hand (manual)
2. Not performing a backup or having a recovery plan
3. Not having a step-by-step plan for designing the integration
4. Not working with IT to setup the network and compute environment before the project

Risk 1: Failing to put systems in hand (manual)

In chapter 13, I discussed the importance of putting systems in hand (also known as manual) control.

Whenever you're performing a systems integration project that could potentially impact existing systems, you will want to put the equipment you are integrating with in hand to ensure that the equipment keeps operating as designed.

I've seen this step missed quite a bit and it almost always results in system failures.

Risk 2: Not performing a backup or having a recovery plan

In chapter 13 I also talked about the importance of backing up your building automation system and having an effective recovery plan. A systems integration project is no different.

You should have a recovery plan laid out for every particular integration that you're going to perform. This recovery plan needs to answer the following questions:

- How to get the system functional again?
- How to rollback any and all changes?
- Who needs to be involved in the recovery process?
- Who needs to be notified during the recovery process?

Risk 3: Not having a step-by-step plan for designing the integration

You would think having a step-by-step plan would be a no-brainer.

However, I've seen quite a few folks try to execute systems integration projects without having a plan at all. I will be covering how to execute systems integration projects in my upcoming systems integration course.

Risk 4: Not working with IT to setup the network and compute environment before the project

This is a common mistake for folks who are new to the systems integration.

Often it is assumed that stuff will just work and that all you will have to do is call up the IT group, they'll give you an IP address and you'll be on your way.

However, this is often not the case when it comes to systems integration. In the previous section, I had you detail out the system requirements as part of your system integration project. Some of you may have been thinking that that was a waste of time.

Now you see why I had you do that. When you meet with the IT group before the project begins, you can detail out all of the technology requirements you have.

That way if you have a requirement that requires a network change or some custom application, your IT group can work with you before the system integration project beginning.

How can you support an integrated solution after it's installed?

Even the best integrations will tend to break over time.

Surprisingly, the biggest problem I have encountered with systems integration projects is not the projects. Rather it is the support of the integrated systems after the install.

I've worked on countless projects, where the lack of a clear and detailed support plan causes the integration to fail over time.

The result of this is that the integration(s), is often overridden, which ultimately puts the facility in a worse position then if they hadn't performed the integration in the first place.

That's why in this final section I am going to give you my 4 step process for creating an effective systems integration maintenance plan.

Here are the four steps:

1. Schedule a reoccurring functionality test
2. Check all software based integrations on a regular basis
3. Make sure that you verify the functionality of the integration after any upgrade or system change
4. Perform regular maintenance on your integration

Step 1: Schedule a reoccurring functionality test

Performing a reoccurring functional test is the first step you need to take to ensure that your integrated systems remain functional after they are installed.

I recommend that you set up a functional test that will reoccur at least twice a year.

Performing a functional test is not very difficult. All you need to do is take your sequence of operations and go through it line by line to ensure that your sequence of operations still functions as designed.

If part of the design does not function properly, then you simply troubleshoot that part of the design utilizing the documentation that was created during the system integration project.

Step 2: Check all software based integrations on a regular basis

It is important that any integrations that are dependent on software are checked at least once a month. To be clear, all you are doing is making sure the integration is working. This is not a functional test.

You are simply checking to make sure that the integration(s) is still communicating and that there are no errors being generated by the integrated systems.

Step 3: Make sure that you verify the functionality of the integration after any upgrade or system change

Anytime an upgrade or system change is made you're going to want to validate that the integrated systems still functions as designed.

Step 4: Perform regular maintenance on your integration(s)

On a regular basis, you will want to perform maintenance on your integrations. The way you will approach this is to check out the physical and logical connections that were defined during the system integration process.

The frequency at which you perform maintenance on your integration(s) will depend on the type of integration.

You will want to validate that each of the physical connections still works. There are multiple different methods that you can use to test if the connection is still functional. The method you chose depends on the type of physical connection.

Once you have completed a check of the physical connections, you will want to verify that all of the logical connections work. You can verify the logical connections by ensuring that data is flowing from one system to another and that any data normalization or storage is taking place.

Chapter 15 quick summary

What a value packed chapter!

In this chapter, I took you on an expedition through the crazy world of systems integration. You learned what systems integration was and how to scope, implement, and support a systems integration project.

In this chapter, I gave you my step-by-step processes to systems integration. These processes alone are worth the price of the book. As with everything in this book, I encourage you to apply what you learned right now!

In the next section, I begin to wrap up the book and show you the next steps you can take to continue your learning journey.

SECTION V – NEXT STEPS AND CONCLUDING THOUGHTS

Section Overview

In this section, I will be wrapping up the book, but don't worry! This is not the end of your building automation journey. In this section, I will teach you how to build a path of continual learning in the best damn career field there is, Building Automation!

CHAPTER 16

Next Steps

What's in This Chapter?

Congratulations!

You've completed the book. The knowledge you've just gained is going to accelerate your career massively. But even more, than that, this knowledge is going to help lift up the Building Automation Profession.

After all, a rising tide lifts up all ships!

If you're like me, you may be wondering what your next steps should be?

Growing your career

Based on my experience, a career in building automation can go down four different paths. My personal career journey flowed through all four of these paths in the order they are listed below.

To be clear, I am not saying you need to follow all four paths. I've met plenty of people who are happily earning high-six-figure salaries who have focused on a specific path.

The four paths are:

- Technical
- Management
- Sales
- Engineering

Technical

For many folks, the technical path is one of the hardest paths to advance in. I believe the reason for this, is that several of the smaller companies are local companies and do not have a path for technical personnel.

The good news is that the technological revolution that is taking place in the BAS world right now is poised to change that.

I've seen and heard from multiple companies that are looking to help their technical folks take their skills to the next level.

What skills should you focus on then?

You could focus on any one of three skills below as you grow your career in the building automation space:

- Information Technology
- Application Development
- Systems Integration

Now before you freak out, don't worry I'm not trying to turn you into an IT professional or application developer. However, I am encouraging you to expose yourself to these skillsets.

In the next section, I will describe how you can do that for each of the three skills.

Information Technology

In regards to information technology, you will want to get to a level where you are comfortable with the operation and design of networks, security, servers, and databases. There are a couple of ways you can get there.

You could go and purchase books on the following topics:

- Network Fundamentals
- Cyber Security
- Client-Server Model
- Database Fundamentals

Or you could attend my course that I will be releasing in 2017 titled, *Information Technology for Building Automation Professionals*.

Application Development

To be clear, I do not expect you to become a developer.

Rather, the important thing is to understand how applications are being developed so that you can work with developers. Controls companies are producing application programming interfaces that allow application developers to access data and create applications.

I would recommend getting books on:

- Use Case Design
- Application Programming Interfaces

In 2017 I will be releasing a video course titled *Application Development Essentials: What a BAS Professional Needs to Know*. In this course, I will describe all of the aspects of an application so that you understand what to look for if you work with a developer.

Systems Integration

Systems Integration is a very popular topic right now, as a matter of fact, outside of the fundamentals of BAS, systems integration is the most commonly requested content on BAM. Based on the way the BAS space is changing, this topic is going to become increasingly critical for you as you move forward in your career.

Management

If you want to focus on the management path and I would encourage you to focus on the following areas:

- Energy Management
- Facilities Operations
- Project Management

Energy Management

Energy management is an area that ebbs and flows. Ultimately facilities need to manage their energy, but the "severity" of the energy management problem depends on the price and supply of energy. In my experience, the quickest way to pick up the knowledge you need is to attend the AEE Certified Energy Manager 5-day boot camp.

I attended this program, and it was invaluable.

Facilities Operations

Managing facilities is a challenge.

How do you create a facility management group?

How do you maintain the training of your building automation technicians?

How do you know when to use a third-party service provider to maintain your building?

These are just a few of the many questions you will need to answer if you want to go down the path of managing a facility, service group, or building automation team.

For facility minded folks, I would recommend checking out BOMA. They have a series of webinars focused on managing facilities.

In 2017, I will be converting this book into an all-encompassing video course titled *Building Automation Systems: A to Z.*

Project Management

Managing building automation projects can be a challenge. However, knowing how to manage a building automation project effectively is an invaluable skill.

That is why I recommend you get training on how to effectively run building automation projects.

In the construction and upgrade chapters, I discussed how projects worked and provided you with some processes that you can use to manage projects. If you plan on focusing on building automation projects, then you should consider getting formal training on project management.

One of the best organizations to learn from is the Project Management Institute (PMI).

The PMI has a Project Management Professional (PMP) certification. While I don't think you need to go to that level, I would encourage you to attend some of the PMP webinars on project management.

You can learn more about PMI here https://www.pmi.org/

Sales

Sales is a challenging place to be right now, the reason for this is that sales roles require a blend of mechanical, BAS, and IT knowledge. After reading this course, you should have the fundamental knowledge around BAS, IT, and mechanical systems that are required to work in a sales role.

However, even with my very thorough overview of IT, you may still want to increase your knowledge around IT systems.

IT

Information technology is quickly becoming an integral part of building automation systems. This means that IT is going to become a critical part of the BAS sales process. However, the IT needs of a technician are completely different than those of a sales person.

While companies like Cisco have IT training for sales professionals, this training is geared towards technology sales persons. There is no training program specifically geared towards training BAS sales professionals on IT skills.

That is why I am releasing a video series geared towards BAS sales professionals in 2017. This course will be called *Information Technology for BAS Sales Professionals*.

Engineering

Design engineers and consultants probably have the largest challenge out of all of the roles I have described so far. Now, don't read into what I said. I didn't say that being a technician, manager, or sales person is not hard.

Rather, what I am saying, is that design engineers and consultants have the largest challenge going forward because of all of the systems that are available to the market.

Imagine you are a design engineer or consultant and your customer is expecting you to be familiar with all of the solutions on the market.

How do you keep up?

For the folks in the design engineering and consulting world, I would suggest that you focus on Information Technology.

I am finding that the IT skills required to design and consult on the newer BAS systems are not being taught by traditional engineering schools and organizations.

Information Technology

Man, this sure is a reoccurring theme, isn't it? I recommended IT training for almost every career path and for a good reason, smart building systems are going IP!

This is why you must invest in your IT learning. If you are a system designer or consulting engineer, you need to seek IT training beyond the scope of this book.

You should understand things like the impact of designing a BAS network that shares the same physical trunks as a video surveillance system...

As field controllers and sensors become increasingly IP-based and control systems begin to support native analytics you are going to have to understand how to design these systems.

There is no IT design course for BAS system designers or consulting engineers. At least not yet.

In 2017, I will be releasing my video course *Information Technology for Building Automation Professionals*. This course will teach you what you need to know to be successful.

Conclusion

That's it folks, the shows over. Or is it?

If you are ready to get your learn on, I encourage you to sign up to the special Building Automation Monthly customers only Facebook Group at

https://www.facebook.com/groups/BAMCustomers/

If you have any questions about the book, please e-mail me at Phil@PhilZito.com

And once again, thank you for being a Building Automation Monthly customer.

CPSIA information can be obtained
at www.ICGtesting.com
Printed in the USA
LVHW051720080120
642937LV00009B/288/P